ISDN Design

ISDN Design
A Practical Approach

Steve Hardwick

Applications Manager
Siemens, Inc.
Santa Clara, California

ACADEMIC PRESS, INC.
Harcourt Brace Jovanovich, Publishers
San Diego   New York   Berkeley
Boston   London   Sydney   Tokyo   Toronto

Academic Press, Inc.
San Diego, California 92101

United Kingdom Edition published by
Academic Press Limited
24–28 Oval Road, London NW1 7DX

Library of Congress Cataloging-in-Publication Data

Hardwick, Steve
    ISDN design : a practical approach / Steve Hardwick.
        p.   cm.
    Includes index.
    ISBN   0-12-324970-8
    1. Integrated services digital networks. I. Title.
    TK5103.7.H37   1989
    004.6--dc20
                                                        89-31226
                                                            CIP

Printed in the United States of America
89   90   91   92   9   8   7   6   5   4   3   2   1

To my wife, Noreen,
with love and thanks for her support.

Contents _____

Preface

With its ever increasing popularity, many engineers are finding that ISDN is affecting their design environment. Unfortunately, ISDN technology covers a wide spectrum of engineering disciplines. Suddenly telephone designers are faced with complex microprocessor systems and terminal designers with telephony. This forces an ISDN designer to understand the other half of the digital network, which until now has only been briefly considered.

This book will give both the newcomer and the experienced design engineer an easy-to-understand reference for ISDN. Topics relating to both theoretical and practical issues concerning ISDN designers are covered. Although some theoretical topics are discussed, it is not the intent of this book to be a definitive technical reference. The object is to use technical information only when it is needed to understand a specific example. Many design examples are given throughout the book.

To begin, there is a brief discussion of the need for ISDN technology, including not only needs from the point of view of telephone design, but also from the perspective of PBX design. Once the outline of ISDN has been given, a refresher of basic concepts needed to understand the subject follows. This includes topics ranging from the theoretical side of data transmission to the requirements of ISDN circuit layout. After this foundation is laid, an overview of the relevant international and national standards that exist in the ISDN telephone world is given in Chapter 3.

Chapter 4 goes into ISDN terminal equipment design. This section is only concerned with issues concerning hardware design. Several examples of circuit designs are given. A wide variety of ISDN terminal design issues are explained with references to these specific applications circuits. The other end of the network, i.e. the exchange, is discussed in Chapter 5. Again, many examples are given for both specific applications circuits and system architectures.

The area of the primary access side of the network, from its place in the system to specific applications examples, is the subject of Chapter 6. The main emphasis is placed on the American system, although some mention is made of European standards.

Chapter 7 reviews the software issues raised by ISDN. These are dealt with from the perspective of the effects upon system and hardware design. Chapter 8 explores the topic of testing ISDN designs, with particular attention given to unique problems facing ISDN designers. Finally, Chapter 9 brings together all of the salient issues facing the ISDN designer and discusses an actual ISDN design.

Because of the complexity of ISDN designs and the relative infancy of the technology, it is very difficult to correlate all the required data. By covering the subject at both a technical and a practical level, an engineer can easily have an appreciation of what ISDN is and how to make a product that will work. In addition, the book also serves as an instructional guide to accelerate ISDN equipment designs.

Acknowledgments

Certain materials were adapted and reproduced with permission of the following companies:

Advanced Micro Devices, Austin, Texas (Figure 4.11b from Am79C401 data sheet and Figure 7.4 from AmLink technical reference manual)

Dayton Development Center, NCR Dayton, Ohio (Figure 9.1)

Siemens Components, Inc., Santa Clara, California (Figures 2.1, 3.12, 4.12, and 6.2a from Siemens ISDN presentations; Figure 4.1 from the IOM rev. 2 specification; Figure 4.2 from the PEB2080 SBC data sheet; Figure 4.11a from the PEB2110 ITAC data sheet)

Tekelek, Calabasas, California (Figure 8.4 from Chameleon 32 product description)

Steve Hardwick

1

Introduction to ISDN

The Existing Telephone Network

The original goal of the telephone, as the name literally suggests, was to transfer "sound over a distance" (voice information). The telephone network that has evolved over the past 90 years or so has been dedicated to do just that by tailoring technology to meet demand. The telephone has operated successfully in basically the same manner for a number of years. There has been no need for technological change as there have been no new telephone requirements. Why then will there be a need for integrated service digital networks (ISDN)?

One reason is the changing perception of what a telephone should do. More and more features have been added to the functionality of the telephone by the *private address branch exchange* (PABX). Many additional operations can be carried out by the telephone in these systems, such as call forwarding, voice mail, and speed dialing. The public network is now seeing an acceptance of these types of services. In fact, these services are now being added due to users' demands. To implement them requires digital equipment.

The early methods of providing such functions used *dual tone multifrequency* (DTMF) signaling. With this technique digital information, such as dialing, could be passed from the telephone to the exchange as a kind of audio signal. This signaling has been enhanced to provide access to various services in the exchange, such as call forwarding, routing to a long-distance carrier, and even the ability to replay messages on an answering machine. Even though this method has gone some way to providing some PABX services to the telephone user, DTMF is not suitable to implement the newer ones.

The original goal of DTMF was to encode dialing from a telephone keypad.

1

Telephone users are now demanding more applications that use alphanumeric data, such as a "digital telephone book" or banking by phone. Because DTMF is not a good method for encoding and transferring alphanumeric data, something else is needed to implement these types of applications.

The second area of new user requirements is personal computer, or PC, data. At first, users were quite content to have a PC sitting at their fingertips to process information. But soon there was a need to transfer information stored on one PC to another. The *local area network* (LAN) market grew to fill this need initially, offering PC users the opportunity to connect computers together in a "local" environment. Thus began the evolution of the data network.

Many of the LAN solutions require specialized wiring, presenting two disadvantages. One is the additional cost of providing the wiring to connect the PCs and the other is the fact that the PC is no longer portable. If a PC user changes location, there is a long wait before the PC follows as part of the LAN. The networking problem does not stop there because many LANs need special equipment to add more computers (nodes) to the network. The PC user often has to have the abilities of a data communications expert just to add a computer to the network.

The use of LANs leads to two networks, one for voice and one for data, which are installed in parallel (Figure 1.1). Redundancy results because two sets of wiring and two separate sets of equipment are used to perform identical functions.

The problem became even more acute when the networks were used to communicate outside a building, or off premise, in a *wide area network* (WAN). Voice WANs have existed for quite some time. Anyone with a telephone can make a connection to many places in the world. The problems of communicating between various locations and different pieces of voice equipment has largely been solved. This is not entirely true for data equipment. One way to get around this problem is to take advantage of the voice WAN by making data look like analog information, that is, voice signals. This is the function of the modem.

Modems have been used for many years. The technology has provided a stable and effective method for allowing data to be transferred across the telephone network. However, there are some fundamental difficulties of this approach, the major one being the bandwidth of the telephone line.

Telephone systems are limited in bandwidth to a range between 200 and 3,400 Hz (see Figure 1.2). This limitation filters out all unwanted signals on the telephone line. Low-frequency sources, such as power line (60 Hz) frequencies, are rejected at the low end of the band pass. High-frequency interference, such as electrical machine noise, is filtered out at the top end. This bandlimiting has a drastic effect upon the speeds at which data can be transferred across the telephone network. The normal way in which a modem operates is to encode the binary "ones" and "zeros" as two different frequencies. These frequencies are attenuated by the telephone network just as any other analog signal would be. As the frequency exceeds the upper band limit of 3,400 Hz, it becomes increasingly more attenuated. Because the time to transmit one bit of information is inversely proportional to the frequency, the rate of data transmission is affected by this bandwidth limitation. If two frequencies of 3,000 Hz and 3,200 Hz were chosen

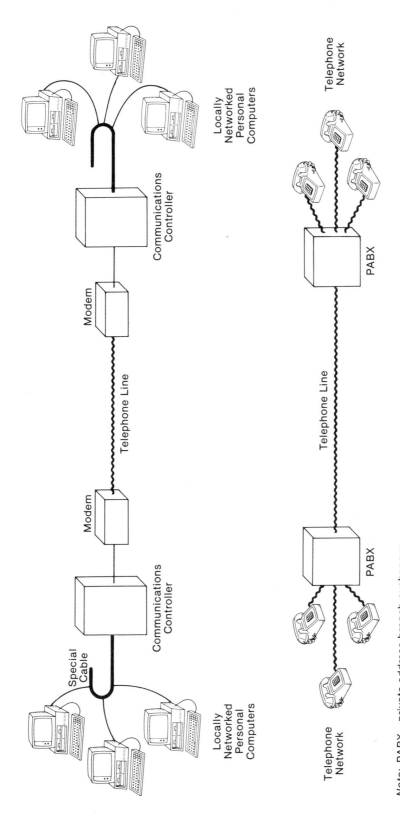

Note: PABX—private address branch exchanges

Figure 1.1. Parallel Voice and Data Networks.

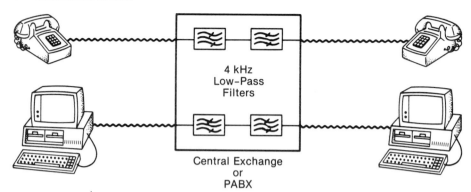

Figure 1.2.   Bandlimiting in the Analog Telephone System.

to represent the data, the maximum data rate would be limited to 3,000 bits per second (bps or bs).

To get a greater data rate, different solutions have been used—for instance using encoding to allow more than one bit to be represented by one single pulse shape or signal level. These solutions require complex circuitry to interface with the telephone network in order to send and transmit the information. By far the best solution would be to treat voice as a data signal, that is, digitize the voice, instead of treating data as a voice signal. Hence, the concept of a *digital network* for the telephone network evolved.

Digitizing voice signals has other advantages too. Currently many telephone exchanges digitize voice signals to allow easier switching of calls. In fact, digitized voice has been used since the 1960s as a method of transferring multiple telephone calls between exchanges. This is known as the *T1 transmission system*. This system can transmit and receive up to 24 telephone calls over specialized telephone lines. By increasing the number of calls over one wire, less wires are needed between exchanges.

Digitization of the network has offered other advantages. By using a digital switching technique in the telephone exchange, many PABX-type services can be offered to the subscriber. The telephone company also reaps the advantage of the digital transition by making it easier, for example, to keep track of billing information. It is no surprise that a great part of the telephone network is already digital.

Unfortunately, telephones themselves are not digital. The main goal of the telephone industry was to support the installed base of equipment that currently existed in the network. There are a lot of telephones out in the network, so the main requirement was to allow change—with the proviso that it would not affect the telephones currently in use. Although much of the network is digital, any change must maintain "plain old-fashioned telephone service" (POTS) to the subscriber. This has been a severe limitation to digitizing the whole network (Figure 1.3).

Increasing pressure has been applied by subscribers to add new services to the telephone network. The penetration of the PC into the consumer market has accelerated the demand for data services. This demand for data services has forced telephone companies to reassess the requirements of the telephone network. This has worked to the benefit of the equipment providers. The telephone

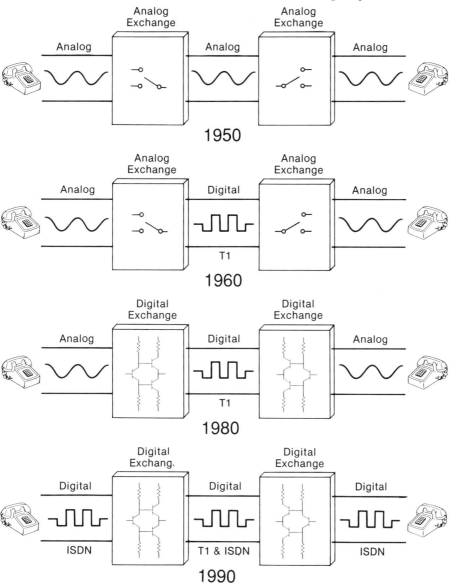

Figure 1.3.   Evolution of the Digital Network.

company can generate more revenue by offering newer and more attractive services.

The challenge to support subscriber demands on the telephone network is not solely a question of providing services. The other side of the coin is standardization. The difficulty of networking various pieces of equipment is not just one of connectivity: customer premise equipment (CPE) has to speak the same "language." For instance, it is easy for someone in the United States to dial someone in Japan. Whether they would be able to have a meaningful conversation would be independent, to a great extent, of the telephone line. The two participants in this case would either have to speak the same language, or have an interpreter. The same is true in data applications.

Maintaining the integrity of digitized voice information is not of great concern. If a few bytes of digitized voice are lost once or twice a second, the end users are unlikely to notice. This is not the same for digital data, in which error-free transmission is a necessity. To overcome this, error detection and correction systems are used. It is very important that both terminals use the same error detection and correction scheme. If not, then the two pieces of equipment are not speaking the same language. For data requirements, there is a need for some sort of controller to be present in the telephone set. This controller helps perform the error detection and correction scheme.

The other function of a telephone set is to provide signaling information to the central office (CO) or PABX. In an analog network, this is done by multiplexing different signals onto the telephone line. For example, the off-hook state is detected by a change in DC current flow on the line; ringing is detected at the telephone by the presence of an AC signal; and dialed digits are encoded into a DTMF format for transmission. This type of signaling cannot be used on a digital telephone line because it would affect the transmission of information. Instead, the telephone and the CO or PABX now have to communicate signaling information in a digital manner. To do this a telephone must have computing capability.

With the need for a computer in the telephone, there is the inevitable need for software to accompany it. Later I will discuss the necessary software for ISDN applications. However, it should be pointed out that the software required for even POTS is considerable. A telephone will be transformed from a simple machine into a small computer, bringing new demands upon the designer. It will be the job of that designer to build a product that will resemble a PC more than a telephone. What are the challenges that face the ISDN telephone designer?

The main difference that the designer will have to deal with is the diversity of engineering disciplines involved in ISDN. Because ISDN is the integration of the data and voice networks, both technical areas are involved. From the telephone world come such elements as voice transmission standards, line theory, and digital switching. From the data world come error correction algorithms, protocols, and packet switched data. The confluence of these two worlds in ISDN applications demands a fundamental understanding of all these topics. Moreover, the telephone is becoming an extension of the PC in many ISDN applications. This means the designer must have some knowledge of PCs and the end-user applications to successfully design an ISDN product that can be integrated into this computer environment and be able to compliment current applications on end-user equipment.

Even a simple POTS design will have new ISDN features. These could include data services in the form of rate adaption or packet switched data; control of equipment in the home (such as heating); and new applications in voice mail. There will also be a market in upgrading existing products, such as facsimile transmission and modem technology. Computers will be affected, as well, with the possibility of the addition of a telephone. This will give rise to a whole new field of applications.

Integrating mixed digital and analog systems will be substantial challenge to the ISDN designer. It is one thing to design an ISDN telephone or voice/data computer; however, it is another to manufacture, test, and produce such a product. The problem of designing a high-frequency digital microprocessor system is almost diametrically opposed to that of designing a high-performance analog telephone circuit. Such problems as layout and power supply requirements will have to be solved.

At the other end of the telephone line—the exchange—there also will be changes. Although many exchanges are "digital," this normally refers to the switching portion only. The telephone line interface is still analog. New telephone line interfaces, or line cards, will have to be developed for the ISDN. The new digital requirements of call establishment that ISDN imposes must now be handled by the exchange, offering new challenges to the exchange system designer. For example, where will the call processing take place? Will it be the responsibility of the line card or will the processing be done as a central function? Or will it be a combination of the two? In any case, new processing power will have to be added to the exchange either at the line card, centrally, or both.

Changes will be seen in the design of small PABXs. No longer will it be possible to construct small switches using analog technology. ISDN will demand that even the smallest switch be digital. The impact on the small switch market will be replacement of small analog PABXs with small digital PABXs for ISDN applications, resulting in increased software overhead. The benefit for end users is increased functionality in the form of more features. One such new feature for small or large PABXs and COs will be data services. The technology of ISDN allows a data call to be handled in the same manner as a telephone call; ISDN exchanges will connect two computers together as easily as making a voice connection. For data, this connection can be shared between computers to effectively use the data bandwidth offered by ISDN. Portions of a data file, or *packets* are transferred discretely over the link. An address is added to the packet to determine its final destination. The network can then route the data packet from the information contained in the address. This is known as *packet switched data*. As ISDN networks grow, the demand to carry this type of traffic will increase. A single ISDN data link of 64 kbs can be used to transmit data from many 2,400-baud terminals in a packet switched connection.

Voice mail services can be added to the switch because voice information in a digital format can be treated similarly to data. A voice message can be stored in the same way as a data file in a computer, offering the advantage that the information can be easily recalled. This service can either be offered at the telephone or inside a switch and will necessitate new planning for PABX and CO system designers, who will have to incorporate storage media for digitized voice information.

The changes ISDN will bring to telecommunications are far-reaching. New engineering skills will be needed to design equipment for the ISDN network. New applications will be spawned by the demands of end users. This book will address the topics that affect the ISDN designer.

2
Basic Concepts of ISDN

Digital Transmission

The *D* in ISDN stands for digital. The first question that immediately springs to mind is, "What does *digital* mean?" Before answering, it is worth looking at the analog network. In analog networks, sound information entering the mouthpiece of a telephone is converted into an electrical signal as long as there is an audio input. The resulting output is a continuous waveform. In simple terms, the electrical signal will vary continuously in response to the sound input. If a sound input varying in a sinusoidal manner were input into the mouthpiece, then a sinusoidal electrical waveform would be output. In a digital system, however, the sound input is sampled at discrete instances in time and output as an electrical waveform. So in the example of a sinusoidal sound input being presented to the mouthpiece, a series of samples of the sound level at various points in time would be the resultant electrical signal.

Methods of sampling the incoming signal to produce a digital waveform have been in use for quite some time. Many publications are available to outline the operation theory of such systems. In this book, these theories will be used but the proofs assumed.

In the previous chapter, reference was made to the bandwidth of the analog telephone system, that is, 200 Hz to 3,400 Hz. The digital network must also handle the same range of frequencies. Using *Nyquist sampling theory*, the number of samples must be twice the frequency of the sampled signal.[1] The sample rate of the analog signal in the telephone system must be at least 6,800 times per second. When the signal is reconstituted, the output is passed through a low-pass filter to remove the unwanted signal components. Because practical filters have physical limitations, a theoretical, sharp cutoff is not realizable. *Aliasing* effects due to the unwanted frequency components from the sampling process occur. To reduce this effect, a guard band is introduced by sampling the incoming signal at a higher frequency, increasing the effectivity of the low-pass filter. In fact, a sample frequency of 8,000 samples per second is chosen. If the sampling frequency is too low, the analog signal is distorted when it is converted from the sampled form. If the sample rate is too high, the bandwidth of the system is underutilized. Once the continuous signal has been converted into a series of sampled levels, these levels are represented by a numerical value. This is the second part of the digitization process.

In the case of ISDN, the sampled levels are measured on a scale from -128 to $+128$. This requires an 8-bit binary value to be assigned to each sample level. Because the sample time for the analog signal is 8,000 times per second, the 8-bit binary numbers that represent these samples are produced 8,000 times a second, resulting in 64,000 binary digits per second. The figure of 64,000 bits per second, or 64 kbs^{-1}, is used extensively in the digital world. It is so much of an axiom that it is referred to as *digital signal* 0, or more commonly DS0. Frequent reference will be made to DS0, or 64 kbs^{-1}, when discussing the digital network.

Once the analog signal has been converted into a series of 8-bit binary sequences, or words, the next stage is to transmit the information over a telephone line. The most common transmission method is to translate the binary digits into pulses, which are then transmitted across the telephone network. Various approaches to the problem of translation will be discussed later in this chapter. The information that is transmitted is digital in nature, and less susceptible to noise signals. Removal of the noise-limiting low-pass filters in the analog network substantially increases the bandwidth. Transmission speed capability of 64 kbs^{-1} is attainable on existing wiring.

In all cases it is necessary to isolate the transmitter and receiver from the telephone line with a transformer. This transformer removes any DC offsets present on the line that might affect the transmission of the pulses. In addition to removing the DC offset with a transformer, it is also necessary to terminate the transmission line. Termination of the transmission line ensures that there is maximum power transfer from the transmitter to the line, and it also reduces the effect of noise on the transmitted signal. Finally, correct termination of the transmission line will reduce the effect of echoes from the line. An incorrectly terminated line will cause a portion of the transmitted signal to be reflected back down the line and cause interference.

Eye Patterns

The stream of binary digits derived from the sampled analog signal is converted to a series of pulses, then transmitted on the telephone line. Even though the 4 kHz low-pass filter has been eliminated from the network, there is still the effect of the transmission line. A telephone line has three elements that determine its characteristics.[2,3] First, the wire itself will have a DC resistance. Second, the wire will also have a self-inductance value that is determined by the geometrical characteristics of the telephone line. Third, there will be a distributed capacitance between the wires of the telephone line. These three elements combine to form a low-pass filter. This low-pass filter will have a significant effect on the transmitted pulse shape (see Figure 2.1).

As can be seen in Figure 2.1, the received pulse has a different shape than the rectangular transmitted pulse shape. If these received pulses are sampled on an oscilloscope at the transmission frequency, then each individually received pulse can be displayed. If the consecutive pulses are superimposed, say with a storage oscilloscope, a pattern will emerge. This pattern, for the pulse shape shown, will resemble a human eye. This pattern is called the *eye diagram*. With this diagram, the performance at the end of the transmission line can be shown and the required performance parameters of the receiver determined.[4]

The first deduction that can be made is that the best sample point is not in the center of the sample period, as might be expected. Because of the time required to charge the transmission line, the point at which the received pulse is the

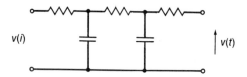

(a) High Frequency Approximation of a Transmission Line

$$v(t) = (1 - e^{-t/T})\, v(i), \text{ where } T \text{ is the time}$$
constant of the line.

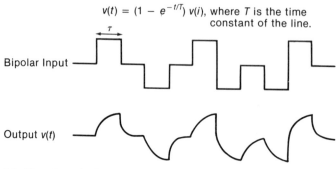

Bipolar Input

Output $v(t)$

(b) Time Domain Response

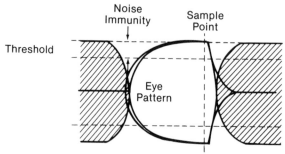

Noise
Immunity

Sample
Point

Threshold

Eye
Pattern

(c)   Superposition of Output Wave Forms

Figure 2.1. Eye Diagram.

greatest is at about 66% of the pulse period. This is known as the point where the eye is the most "open." Choosing this sample point maximizes the system's noise immunity. Therefore, for any given system, two conditions can be calculated: the eye diagram for zero transmission line length and that for maximum transmission line length. By looking at these diagrams the best point for sampling the incoming waveform can be found.

The eye diagram can also be used for giving a simple visual representation of the effect of other system conditions. For example, if noise is present on the line, it would be superimposed on the pattern and cause a reduction in amplitude, or closing, of the eye diagram. By including the sample point and the receiver sampling threshold level, the maximum amount of receiver noise tolerance can be found.

The eye diagram can be used to calculate the receiver requirements. Unfortunately, it is not practical to use the eye diagram to specify the output of the transmitter. Instead, the input pulse needed to generate a specific eye pattern at the end of the transmission line is calculated. This calculated transmitted pulse shape can then be used to specify the transmitter output requirements. In a practical application, there are tolerance limits of the transmission medium and the receiver. To accommodate these, two ideal pulses are calculated; the actual output from the transmitter can lie between these extremes. These two boundary conditions form a *template* for the output pulse. The template in effect "guarantees" the receiver eye pattern limits. As we will see in the next section, the eye diagram has other important uses in evaluating the effect of system parameters on overall performance.

Jitter

In any digital transmission system, it is essential that the received pulse stream be sampled at the same frequency as that of the transmitter that generates the pulses. If the sample frequencies are not matched, errors will occur in the system.[5] In the case of digitized voice, this will result in noise and signal distortion. The sampling rates at the receiver and the transmitter must be synchronized. Some transmission systems use extra lines to provide a synchronization signal from the transmitter to the receiver. These systems are known as *synchronous systems*. In many cases the synchronization information is extracted from the incoming pulses, thus reducing the number of transmission lines. These systems are known as *asynchronous systems*. For example, four wires are required to carry one transmit pulse stream, one receive pulse stream, and clocks to give the pulse period and the start of the analog signal sample period. If only data are transferred and the clock information extracted from these pulses, then only two wires are needed. This results in a saving of 50% of the wiring of a synchronous system.

The clocking information is normally derived by means of a *phase-locked loop* (PLL) connected to the receiver. The PLL will generate a clock from a high-frequency oscillator. The frequency of this clock will be altered dynamically to track the frequency of the incoming pulse stream. This will involve altering the pulse width of the clock generated by the PLL. This variation of the clock period is called *jitter*. Another way of viewing jitter is to consider it a form of frequency modulation. The ideal clock is the carrier waveform; the variation of the period by the PLL to track the incoming pulse stream is the modulating frequency (see Figure 2.2).

Jitter can be specified as an analog signal. Its minimum and maximum frequencies can be defined. In fact, low-frequency jitter is often referred to as *wander*. The amplitude is normally measured as a percentage of the nominal pulse width. This type of jitter is sometimes specified by referring to the waveform shape of the jitter signal. The common type is *sinusoidal jitter*, in which the jitter signal is a sine wave of either fixed or varying frequency. Because the jitter is a result of the PLL attempting to synchronize a clock to an input pulse stream, the input data stream is often specified. Data streams of all 1s or all 0s are used or a sequence of pseudorandom values can be specified.

The effect of jitter on receiver performance can be shown by using the eye diagram. Assume one of the functions of the receiver in the terminal is to extract clocking information from incoming data using a PLL. The incoming pulse is sampled at a defined time from the zero crossing. If the jitter is too great, the PLL will have difficulty in adjusting to the deviation from the nominal clock period. A limit will be reached at which the PLL cannot track the deviation. Therefore, the frequency and amplitude of the jitter signal have an effect upon the sampling of the incoming data. Too high a frequency or too great an amplitude will cause the

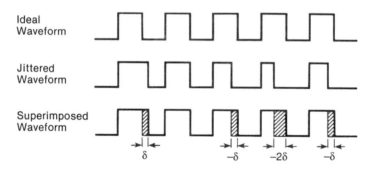

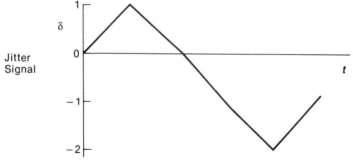

Figure 2.2. Jitter Waveform Representation.

incoming data to be sampled incorrectly, resulting in bit errors on the transmission. These errors will cause incorrect transfer of data or added analog signal noise to a voice transmission.

Coding Schemes

Another aspect of digital transmission systems is the algorithm used for data transmission on the line. There are various ways in which the data can be represented as signals on the line. Before discussing this, it is worth pointing out the difference between *bits per second* (bs) and *baud rate*. The number of bits per second gives an indication of the number of binary digits that are transmitted in a second, that is, data per unit of time. The baud rate however, is a measure of the maximum number of sample intervals per unit time. In many cases the baud rate and the bit rate are the same value but in certain cases the baud rate and the bit rate are different. The latter type of encoding is used so that more bits can be transmitted at a given frequency. The result of this increased data rate is a more complex transmitter and receiver. One of the important factors of a transmission system is the number of discrete levels a signal can have.

For example, TTL has two levels to encode the binary data. One voltage level signifies a logic 1, and the other voltage level a logic 0. A single switching point, or *decision level*, is used to set the threshold. By comparing the coded signal to the threshold level, the binary information can be decoded. If the signal level is above the threshold, then a logic 1 is detected; conversely, if it is lower, a logic 0 is detected. Because the signal has two levels, it is known as a *unipolar code*. Unipolar coding is a very good system for TTL systems but is not suitable for transmission systems using copper cables over long distances.

The reason why lies in the construction and characteristics of such cables. A transmission cable is made of a group of wires bound in a casing. The wire itself has two electrical properties. One is the DC resistance (ohmic resistance) and the second is the self-inductance value of the wire. When a group of wires is placed in the casing, a third electrical property, capacitance between the wires, is brought to bear. The transmission cable looks, in effect, the same as a low-pass filter. If a unipolar code is transmitted down the line, each high-level signal will inject energy into the capacitor/inductor of the cable. Conversely, a low-level signal will discharge the line. If the number of highs and lows is matched, then the net energy level on the line will be zero. But if the number is not matched (which is normally the case), the transmission line will have periods during which energy is stored. This stored energy will result in a DC offset being superimposed on the transmitted signal. The DC offset will interfere with the decoding of the received pulse by reducing the difference between the high and low signal levels and the threshold level. For instance, if the threshold level is 1.3 V and a low-level signal is 0.8 V, then a DC offset of greater than 0.5 V will stop the decoding of a low-level signal and affect the overall performance of the transmission system.

To overcome this, transmission codes with three levels, *positive high*, *zero*, and *negative low*, have been developed. The highs and lows are used to represent the same logic level. The main difference between this system and the unipolar system is that the highs and lows are alternated for the same logic level. For example, in the system of coding known as *alternate mark inversion* (AMI), a high or a low level represents the logic level 1. A zero level represents the logic level 0. To transmit the sequence 101 the following pattern would be output: high/zero/low. This type of coding is known as *pseudoternary coding*. It is called "pseudo" because the zero voltage level is not really classified as a discrete level. The receiver needs two decision levels to decode the incoming data and also has to keep track of the level of the last logic 1 transmitted so that the correct level of the next logic 1 is sent as the code alternates. In the case in which the wrong level is detected, that is, a high followed by a high, a code violation is recorded. The advantage of this type of coding is that the DC level, or balance, is maintained by the transmission of alternate high and low pulses. The disadvantage is the extra circuitry required.

Another type of encoding that is commonly used is a *block code*. A block code is used to transmit a high data rate over a limited transmission frequency. In a block code, a sequence of transmitted pulses is used to represent a unique binary number.[6] (See Figure 2.3.)

The reason block codes are used is to realize a higher data rate down the transmission line. As outlined in Chapter 1, the transmission line has some important characteristics, with a major one being attenuation of a transmitted signal in proportion to its frequency. The higher the frequency, the less signal at the receiver for a given distance of cable. Alternatively, the higher the frequency, the shorter the transmission distance for a given receiver input level. By using a block code, the number of bits per sample period can be increased, making the data rate on the line greater than the baud rate. For a given performance level of a receiver, the range at which a certain data rate can be achieved is increased by using a block code. The price paid for increased transmission distance is more complex circuitry due to the addition of an encoder to the transmitter and a corresponding decoder to the receiver.

There is one more trick used by the transmissions systems equipment designer. Many receivers extract clocking information from incoming data by detecting the occurrences of the incoming signal data crossing a decision threshold. Each time the signal crosses this point, a PLL can lock its output to it. To perform this task effectively there have to be sufficient crossings over a period of time. If the number of crossings is reduced, the PLL can drift from the frequency of the incoming signal. This will cause the incoming signal to be sampled incorrectly and result in data errors. With a ternary code, the logic levels encoded into alternate high and low output signals provide the clocking information. For example, in AMI coding, each time a logic 1 is transmitted the receiver can lock the PLL onto the incoming signal. Unfortunately, a sequence of logic 0s transmitted in AMI coding will not provide clocking information as no transitions will occur. To overcome this problem, the practice of *zero code suppression* is used.[7] The date stream to be transmitted is examined to determine if a long sequence of logic 0s is about to be transmitted. If such a sequence is detected, a special pattern is deliberately added to the data after a defined number of logic 0s.

To look at a specific case, *binary 8 zeros suppression* (B8ZS),[8] it may be decided that for a particular system no more than eight logic 0s can be transmit-

4-Bit Binary Code	Code Set 1		Code Set 2		Code Set 3		Code Set 4	
	Ternary Code	Next Set	Ternary Code	Next Set	Ternary Code	Next Set	Ternary Code	Next Set
0000	+ 0 +	3	0 0 −	1	0 − 0	2	0 − 0	3
0001	0 − +	1	0 − +	2	0 − +	3	0 − +	4
0010	+ − 0	1	+ − 0	2	+ − 0	3	+ − 0	4
0011	0 0 +	2	0 0 +	3	0 0 +	4	− − 0	2
0100	− + 0	1	− + 0	2	− + 0	3	− + 0	4
0101	+ + 1	2	+ + −	3	+ − −	2	+ − −	3
0110	− + +	2	− + +	3	− − +	2	− − +	3
0111	− 0 +	1	− 0 +	2	− 0 +	3	− 0 +	4
1000	+ 0 0	2	+ 0 0	3	+ 0 0	4	0 − −	2
1001	+ − +	2	+ − +	3	+ − +	4	− − −	1
1010	+ + −	2	+ + −	3	+ − −	2	+ − −	3
1011	0 + 0	2	0 + 0	3	0 + 0	4	− 0 −	2
1100	+ + +	4	− + −	1	− + −	2	− + −	3
1101	0 + 0	2	0 + 0	3	0 + 0	4	− 0 −	2
1110	0 + −	1	0 + −	2	0 + −	3	0 + −	4
1111	+ + 0	3	0 0 −	1	0 0 −	2	0 0 −	3

Examples

Code for 4 3 2 1

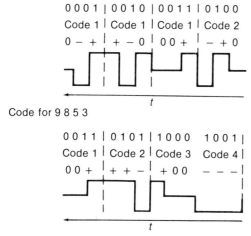

Code for 9 8 5 3

Figure 2.3.   4B3T Code.

ted sequentially. After the occurrence of eight or more logical 0s, a special pattern (000+−0−+ after a preceding high mark, +, or 000−+0+ after a preceding low mark, −) is added to the data stream. This special pattern is unique because of the imbedded code violations. When detected at the receive side it is removed and not seen as a code violation. If a string of nine logic 0s were to be transmitted, then a special pattern would be added to the stream after the eighth logic 0. At the receive end, the special pattern, which is expected by the receiver, is decoded back to eight zeros removed. The data output by the receiver is still the nine logic 0s that were to be transferred.

Two things can be seen from the example. The addition of the extra, special pattern gave the receiver the clocking data it required to allow the PLL to maintain synchronization. Secondly, to transmit the sequence of nine binary digits still requires only eight pulses on the line. Although the receiver will be less likely to lose synchronization, more complicated transmitters and receivers have to be used.

Framed Data Transmission

A fundamental requirement of ISDN is to allow data to be transmitted from point to point and reach a destination without errors. The rate of transmission of the data in the ISDN is 64 kbs^{-1}, so any error must be detected quickly and corrected as rapidly as possible. Additionally, many data interfaces currently operate at a lower frequency than the 64 kbs^{-1}. To effectively use the bandwidth provided by the ISDN, a mechanism to allow data channels to share the link is required. This can be accomplished by using *packetized data transmission*. As the name implies, the data are broken down into small "packets" for transmission. Each packet contains a variable number of data bits that is less than the system maximum. These packets are then transmitted over the transmission line. An algorithm for error detection can be added into the system to ensure the integrity of the data transfer. If the length of the packet is small, that is, if the number of data bits in the packet is low, then any error in transmission can be detected quickly. A request to retransmit the damaged packet of data can be given and the system can recover from the error quickly, thus meeting the requirement to have error-free data transmission.

If a unique address were to be added to the packet, then the information's reception point could be identified. Several transmitting stations could share the same transmission line. The address on the packet would allow the packet to be directed at the end of the transmission line. A good analogy is the mail service. Any person can place a letter in a mailbox and the letter will wend its way to the addressee. The address on the letter allows the postal service to route the letter to its intended destination. By using a similar type of addressing system, several transmitting stations can output data packets, in turn, on the transmission line. Because the data rate of the line is fixed, the sum of the required rates of the transmitting stations must be less than that of the transmission system. The link made between the transmitting station and the receiving station is known as a *virtual connection*.

To implement such a system of packetized data, three criteria must be met:

Existence of a method for "fracturing" the data into packets
An error detection algorithm
Packet addresses to route the data to the correct destination.

A system called *high-level data link control* (HDLC) protocol has been developed to meet the above requirements.

To section the data into packets, a delimiter must be chosen to signify the start and finish of each packet. This delimiter must be a sequence of binary digits that will not appear in the data stream. To perform this task, the binary sequence 0111 1110 has been chosen to delimit the packet boundaries in the HDLC protocol. Under normal circumstances, it is highly probable that this sequence will appear in the transmitted data stream. To remove the occurrence of this event, the data stream is first encoded. A binary 0 is inserted into the data stream when the occurrence of a pattern of six or more logic 1s is detected. This extra digit is removed by the receiver and so is "transparent" to the transmitted data. The increased 0 density aids clock extraction on certain transmission systems where a logic 0 is represented by a pulse output on the line. The price that is paid is reduction of the effective data rate of the system.

To meet the second requirement, a *cyclic redundancy check* (CRC) error detection algorithm is used. It is beyond the scope of this book to go into the detailed math behind this technique, but a CRC is generated by the following process. The number to be checked is first multiplied by the number of bits the CRC contains. For example, for a 16-bit CRC the data is multiplied by 2^{16}. The resulting product is divided by the CRC *generator polynomial*. The polynomial is given in the standard that defines the different types of CRC systems that are in use. The example shown in Figure 2.4 uses the CRC standard of the International Telegraph and Telephone Consultative Committee (CCITT; the abbreviation is derived from the French form of the body's name). The remainder after the division is used as the CRC value. At the receiving end, the number with the last 16 bits set to 0 is divided by the generator polynomial and the remainder compared to the CRC value. If the remainder and the CRC are equal, the received data is considered error free.

CRCs are used in other applications. As a case in point, a CRC is often used to verify data written to a floppy disk. A CRC is a very easy way to implement an error detection scheme with a low overhead of additional data. The divisor used in CCITT protocols is the polynomial $x^{16} + x^{12} + x^5 + 1$, or 1 0001 0000 0010 0001. The division used for generating the CRC is *modulo 2*. In many cases, the overhead to perform this type of calculation at the data rates required by ISDN is too "CPU-intensive" for the task to be executed by the system microprocessor.[9] Instead the task is left to specially designed circuits in communications devices.

The third requirement is an address for the packet. In the case of HDLC, a 2-byte address field, corresponding to two *identifiers*, is used. In effect, each identifier refers to two separate destinations. The first is the *service access point identifier* (SAPI). The SAPI informs the receiving station which service, or part of the software, the packet is destined for. The second is the *terminal endpoint identifier* (TEI). The TEI selects which terminal the packet is destined for. The SAPI is a software address; the TEI is a hardware address. The TEI can be used to send packets between different stations over the same link, giving rise to a virtual link between stations. The SAPI is used to signify the different functions performed over the link. Functions such as management of the link or transmission of data would have different SAPI values.

```
                                                       1000 1010
1 0001 0000 0010 0001    1000 0010 0000 0000 0000 0000
                         1000 1000 0001 0000 1
                         — — — — — — — — —

                         000 1010 0001 0000 10
                         0
                          — — — — — — — — —

                          00 1010 0001 0000 100
                          0
                           — — — — — — — — —

                           0 1010 0001 0000 1000
                           0
                            — — — — — — —

                             1010 0001 0000 1000 0
                             1000 1000 0001 0000 1
                             — — — — — — — —

                             010 1001 0001 1000 10
                             0
                              — — — — — — — —

                             10 1001 0001 1000 100
                             10 0010 0000 0100 001
                              — — — — — — — —

                              0 1011 0001 1100 1010
                              0
                               — — — — — — — —

                               1011 0001 1100 1010
```

Note: CRC for letter A is 538D (the CRC and the binary value for
    A are written LSB first).

Figure 2.4.   Generation of CRC for ASCII Value of Letter A.

One additional feature that has been added to the HDLC protocol is the *control field*. The control field is used to signify different types of packets. There are three major types of packets in HDLC. These are *information* (I) frames, *supervisory* (S) frames, and *unnumbered* (U) frames. One of the stipulations of the HDLC protocol is that every transmission of a frame carrying information should be acknowledged by the receiving station. This acknowledgment is done by using an S frame. For each transmission there is a corresponding acknowledgment frame transmitted from the receiving station.

To ensure that a reasonable amount of data is transmitted before an error check is performed, the maximum frame length is fixed for a system. If the information to be transmitted has more bytes than the maximum frame length allows, then the data must be transmitted in multiple frames. In this case, a record must be kept of the frame number transmitted so the frames are transmitted in sequence. This is even more important if an error is detected in the receive stream. When an error is detected, a request for retransmission of the frame, instead of an acknowledgment, is sent from the receiving station to the transmitting station. It is vitally important that the receiving station know that the frame it receives is the retransmission of the previously corrupted frame. This can be determined by the sequence number of the frame. In this fashion a complete error-free transmission of a block of data can be accomplished.

There is another important parameter in connection with the transmission of HDLC-framed data. Although it is necessary to respond to each frame transmission with an acknowledgment frame from the receiving station, in certain cases this is impractical. In a satellite link for example, the round-trip time would represent an unacceptable delay as the transmitting station waited for an acknowledgment frame. To overcome this, an acknowledgment is generated only after the reception of a certain number of complete frames. The number of frames that can be received before an acknowledgment is sent by the receiving station is called the *window size*.

Analog-to-Digital Conversion and Digital-to-Analog Conversion

One of the steps that has to be provided in ISDN to support voice information is the conversion of the analog voice to binary digital data and vice versa. To perform this function a device, *coder/deco*der (codec), is used to code and decode the voice into digital data. In the ISDN, the sample rate is 8 kHz and the number of bits is eight, resulting in a range of 256 finite levels that the coding system can have.

The incoming analog waveform is sampled at the 8 kHz rate, one sample every 125 microseconds (μs), and then converted into a digital word. The first part of this process is deciding the best level to represent the input signal. Because there are only 256 levels in the digitized analog signal, there will be a range of signal levels that corresponds to the same digital word. If the interval between the levels is N, then the range of signals around a given level will be $\pm N$. In the reverse process, the digital data are converted back to an analog signal. The first step is to convert the digital data into an output level. The second step is to reconstitute the analog signal by passing the output of the decoder through a low-pass filter, or *anti-aliasing filter* (see Figure 2.5).

One of the inherent difficulties of this method is the approximation of the incoming signal by the coder. This will result in a slight difference between the input signal to the digital transmission system and the output from the system. This is known as *quantization distortion* and can be seen in all digitization systems. In the analog telephone network, the range of speech volume varies from user to user, which results in a dynamic range requirement of about 40 dB for the telephone link. If a coding/decoding system has a constant interval between the digitized levels (*linear encoding*), then the quantization distortion between the various users will vary. For example, a user with a large dynamic range would have less distortion than a user with a smaller dynamic range.

To overcome this problem, a scheme can be used in which the interval between the various levels used in the coding/decoding process is not uniform. When the input signal is converted using a system with *nonlinear* levels, this is known as *compression*; the reverse process is called *expansion*. These terms are

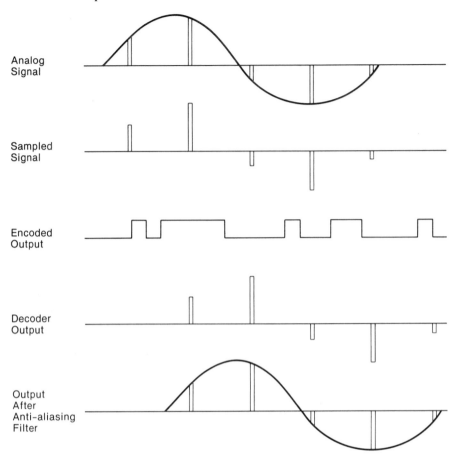

Figure 2.5.   Coding/Decoding.

normally joined together and the method of converting analog signals with a nonlinear coding is called *companding*.[10] Because in analog voice the density of the probability of various amplitudes in the speech input has been qualitatively measured, a logarithmic companding scheme has been found to be the most suitable. This logarithmic scheme reduces the amount of quantization distortion added to the speech through the system.

There are two logarithmic companding schemes currently in use today. One is the *A-law* and the other is the *μ-law*. Although not truly logarithmic companding, both A-law and μ-law are practical realizations of logarithmic companding. The A-law scheme is more commonly found in Europe; the μ-law scheme is more commonly found in North America and Japan. The standards that apply to these methods are covered in chapter 3.

Along with the dynamic requirement of 40 dB for voice band signals, there is also a need for a signal-to-noise ratio of better than 30 dB. Practically, this would require a resolution of around $1:8,000$. A precision analog circuit is needed to achieve this type of performance with a conventional analog-to-digital (A/D) converter. This type of circuit can be difficult to produce as an integrated circuit;

it is often demanding to manufacture close tolerance component values. To achieve the desired performance, one technique that is becoming more frequently used is interpolation.[1]

Interpolation uses a system of conversion that is similar to *delta modulation*. In delta modulation, the analog input is estimated, and this estimation is compared to the original input and increased or decreased accordingly. Interpolation uses the same principle except that the input analog signal is sampled at many times the Nyquist frequency. These samples are then averaged over the Nyquist sampling frequency of the analog input signal. The resulting digital word of 13 to 16 bits is then compressed using an A-law or μ-law compression algorithm. To achieve the required signal-to-noise ratios over the voice dynamic range, the input signal is oversampled at a rate of 256 kHz to 512 kHz, resulting in a 16-bit sample every 125 μs (8,000 times a second). Although the sample rate for this type of conversion is often given as 256 kHz or 512 kHz, the maximum analog frequency that the converter will handle is still 4 kHz.

Pulse-Coded Modulation

Once the voice has been digitized, the resulting digital data can be passed through the system. Part of the process is routing the information to the correct destination. This function is performed by the CO or PABX. In the early exchanges, connection was achieved using mechanical switches to connect the two sets of telephone lines from the subscribers together. As technology evolved, these mechanical switches were replaced with electrical ones. The form of switching was still the same; however, each telephone line required a switching circuit. The central part of the exchange that performs the switching function is normally referred to as the *switching matrix*. It is more and more common that the switching matrix in exchanges are digital.

In the current analog telephone network, incoming telephone signals are received by the exchange and digitized. The digitized data can then be routed by a switching matrix. A coding technique called *pulse-coded modulation* (PCM) is used to convert the analog signals. The data are output in a serial bit stream with a logic 1 being represented by a pulse and a logic 0 represented as a space (no pulse). The exchange will use these digitized data to route the telephone information from one line to the next. One advantage of having the telephone information in this format is that by increasing the bit rate of the data, several telephone lines can be multiplexed together using *time division multiplexing* (TDM).

The standard rate for the digitization of voice is 64 kbs. If the rate at which the data are sent from the coder is increased, then the coder can send its 8 bits at the high frequency and leave a gap between the transmissions. For example, if the bit rate for the digitized voice is 2,048 kbs, then μs transmission will take only 3.9 μs leaving a gap of 121.1 μs before the next 8 bits are sent. By using a bit rate of 2,048 kbs, 32 digitized telephone lines can be multiplexed using TDM. This would allow the 32 telephone calls to be carried inside the exchange over a single set of wires. Because the distances involved inside the exchange are relatively small, it is possible to transmit digitized voice at these rates.

Digitized voice from the 32 telephone calls can be sent to the switching matrix over the same two wires at the 2,048 kbs rate. This substantially reduces the amount of wiring and circuitry required inside the exchange. A codec is now needed for every telephone line coming into the exchange to convert the analog telephone lines to 2,048 kbs multiplexed digital voice signals. These codecs are normally grouped together with the circuitry required to change the data rate on a line card. When the digital data are multiplexed, the 3.9 μs time slot it occupies is called a *PCM time slot* or *channel*. The 32 multiplexed telephone lines that are transferred over the same wiring are called a *PCM highway* (see Figure 2.6).

Inside a typical exchange there will be a PCM highway for every 32 telephone lines. Each highway will consist of a transmit line and a receive line. In addition, a system clock and a frame pulse are provided to all of the line cards in the exchange to allow the digitized voice data to be multiplexed. The frame pulse will be at the same frequency as the sampling rate of the voice, that is, 8 kHz. There are several frequencies used for the data rate of these types of systems. The most popular rate is 2,048 kbs, although with technology becoming more sophisticated, rates of 4,096 kbs and 8,192 kbs are become more widespread.

By using this encoding scheme it is easier to manufacture larger exchanges. TDM reduces the amount of internal wiring inside the exchange. The PCM high-

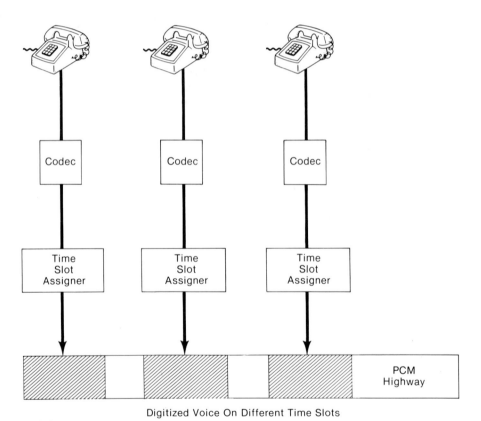

Digitized Voice On Different Time Slots

Figure 2.6.   PCM Highways Used to Multiplex Digitized Voice.

way is the backbone of digital exchanges, and some exchange functions can be put into specialized semiconductor devices. For example, there are many devices available that will interface to several PCM highways and perform the switching function by allowing the 64 kbs time slots to be switched from highway to highway. If one highway is for the incoming telephone lines and another for the outgoing, then these chips will cross-connect the telephone lines on the PCM highway. By using this type of technology, more functionality can be made available. With ISDN, this functionality can be more easily passed on to the subscriber.

Fabrication of ISDN Designs

One of the challenges that ISDN brings is the need to manufacture new equipment, and this in itself will require new expertise. In the current analog environment, the majority of subscriber equipment is made using analog technology. With ISDN, there will be a need to incorporate digital circuitry into such designs. This will bring new design, layout, testing, and fabrication requirements to the manufacturers of telephone equipment.

The first area that will be affected is layout of ISDN designs. The challenge here is to make a cost-effective design that will mix the analog and digital requirements of ISDN. For ISDN to be a lucrative venture, the products must compete with the existing analog equipment. Although many line cards are of a mixed digital and analog design, there are still some new problems that the layout engineer must solve when implementing an ISDN design.

Perhaps the most obvious candidate for layout problems will be a voice/data workstation. Here there will be both sensitive analog circuitry and high-frequency digital circuitry. The performance required by the ISDN must be at least equal to that demanded by the current analog equipment. This translates to a design goal of a low-noise figure for the analog front end. In principle this may sound easy enough to ensure; however, in a practical environment it may not be so easy.

For example, consider a design that will turn a PC into a voice/data workstation. In many cases the PC format allows for external connections to be made from one edge of the card. This will mean that both the high-frequency line interface and the sensitive analog input will be located in close proximity to each other. Care will have to be taken that the noise levels in such a design are kept within acceptable limits. A second example is a line card. Typically a line card will handle eight or more telephone lines. Again, because of space limitations, the line interfaces will be physically close together. This can cause problems with cross talk between adjacent lines. When data are being transferred over the line, errors can be generated because of the cross-talk effect.

One technique that can help to reduce noise and cross-talk effects is the use of a *ground plane* in the layout design. By the proper use of ground planes, two advantages can be gained: the electrical isolation of various parts of the circuitry, and equal paths for the signal and the ground return. By providing individual ground planes for the various parts of the circuit, the effect of the digital circuitry's high-frequency fluctuations can be minimized. Care must also be taken to correctly decouple the power supplies to the various parts of the circuit. Most analog designs require a split rail supply ($\pm$ 12 V) that must be decoupled to the analog ground plane.

By providing an equal path for the analog signal's ground return, the DC offset seen at the analog input is reduced. This reduction will give a better analog performance by attenuating the distortion effect. The return ground signal will be induced into the ground plane following precisely the tracking of the signal connection, ensuring equal path length.

Another area of concern in mixed analog–digital designs is ground paths. By using a ground plane technique, the ground paths can be minimized. This will lessen the possibility of *ground loops*. The analog and digital grounding systems can be connected at a single point, typically at the power supply. This will ensure a high AC impedance while maintaining a low DC impedance. If several separate ground planes are used, then a star connection can be implemented to connect these paths together. Many semiconductor devices that incorporate both analog and digital circuitry (codecs, for example) have both digital and analog ground pins. Care must be taken that the correct layout is used for the connection of these pins on the device.

Multilayer boards provide many advantages in the layout of ISDN equipment. The manufacturing costs of this format are decreasing, and this will encourage large-scale production for use in telephone equipment. This type of technology offers better electrical performance (due to more effective ground and power planes) and a saving of space (due to less board area required for device interconnection). Multilayer boards offer one additional advantage that is particularly useful for mixed analog and digital designs. As the frequency of an electrical impulse increases, the bulk of the conduction is performed at the outer surface of the conductor, and the rest of the conductor acts as an insulator. This is known as the *skin effect*. This effect can be utilized in digital designs. For instance, if the high-frequency signals are routed on the layer above a 5 V plane on a multilayer design, then the induced voltage in the plane will only be on the surface of the signal digital path. The rest of the plane will act as an insulator to the signal. By routing as many high-frequency signals on the layer above the + 5 V power plane as possible, the coupling into the ground plane will be reduced. Because many board layouts are routed using an algorithm of alternate layers carrying orthogonal signal tracks, some care should be taken to place the high-frequency signals on the power plane side of the layout.

A further step that can be used to decouple the digital and analog sections is to separate the areas within the ground plane itself. Isolation boundaries can be etched into the copper ground plane to separate the various ground systems. In an ISDN design, it may be necessary to have more than two ground plane systems to ensure maximum electrical separation of the various parts of the circuit. For example, one is needed for the analog telephone interface, a second for the line interface, a third for the clock generation circuitry, and a fourth for the digital circuitry. By carefully routing the digital signals, which may be in the order of 10s of megahertz, noise levels of better than 70 dB may be achieved (see Figure 2.7).

The connections to the analog input should be kept as short as possible. If there are long connection paths to the analog inputs, these can act as antennae and receive the high-frequency emissions from the digital circuitry. Considera-

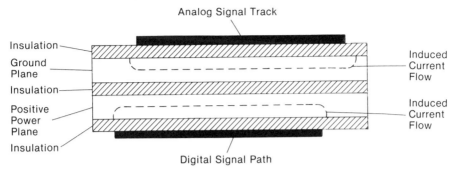

Figure 2.7.   Effects of Multilayer Planes.

tion should be given to inductive coupling of high-frequency signal paths to analog input paths. To take the previous example of a plug-in card for a PC, the signal paths for the analog inputs should not be made parallel to those of the line interface. This could cause inductive coupling of the line output signal to the analog inputs. This is equally the case for the line card, in which there is a tendency to "step" the layout design. This could cause inductive coupling between the interfaces and increase the effect of cross talk.

Layout will not be the only area affected by the switch to ISDN. Testing of ISDN designs will be a problem because of the amount of system integration performed at both the terminal and line car. For example, before any performance of the telephone channel can be measured, the line interface must be activated so that the channel can be allocated. Some form of simulation of the corresponding part of the network must be implemented to facilitate testing—for example, a network termination must be available to rest terminal equipment and vice versa. Additionally, both analog and digital tests are required. This poses the problem of developing test philosophies that will yield a cost-effective product. In the design phase, attention should be given to a final product that is easily testable (see Chapter 8).

Units and Terminology

A new area for many designers of ISDN equipment is that of units of measurement. The design engineer has much to learn in this area. Many of the standards that are used for ISDN originated in the analog telecommunications world. This also means that though many of the standards are international, some of the units are not. Take, for example, analog signal levels. Noise levels in particular are related to the system of compression that is used for voice signals. Thus, A-law encoding has one set of units whereas μ-law encoding has another. In chapter 3 (on standards) the difference between these measurement systems will be explained.

A particular quantity can be measured in different ways. With each method of measurement there is an associated set of units. For example, jitter can be measured as an amplitude in which the measurement is given in percent pulse width or unit intervals. Similarly, the input jitter to a system is specified as a frequency plot of the jitter modulation signal. There is also the difficulty of working with different engineering disciplines. For example, noise in the telecom world is specified in decibels, whereas in the digital world it is usually expressed as microvolts per volt.

The method in which specifications are presented may be confusing to designers unfamiliar with the terminology of the different spheres of the engineering world. Line output specifications are given both as a discrete parameter, in the case of frequency, and also as a template. In the data world there are various methods to measure the transfer rate of the data over a connection, such as line rate, effective data rate, and baud rate. It is very important to understand these units when designing and specifying ISDN equipment.

References

1. Bell Labs, "Transmission Systems for Communications," pp. 110–14.
2. Eugene Riley and Victor Acuna, "Primary Parameters of Cable Pairs," in *ABC of the Telephone*, vol. 7.
3. Eugene Riley and Victor Acuna, "Secondary Parameters of Cable Pairs," in *ABC of the Telephone*, vol. 7.
4. Bell Labs, "Transmission Systems for Communications," pp. 708–14.
5. Bell Labs, "Transmission Systems for Communications," pp. 733–41.
6. Siemens, "PEB2090 ISDN Echo Cancellation Circuit Data Sheet."
7. Bell Labs, "Transmission Systems for Communications," p. 745.
8. William Flanagan, "The Teleconnect Guide to T1 Networking," Telecom Library, Inc., pp. 52–53.
9. Greg Morse, "Calculating CRCs by Bits and Bytes," *BYTE Magazine* (September 1986): 115–24.
10. Bell Labs, "Transmission Systems for Communications," pp. 616–28.
11. James C. Candy, William H. Ninke, and Bruce A. Wooley, "A Per-Channel A/D Converter Having 15-Segment μ-255 Companding," *IEEE Transactions on Communications* vol. com.-24, no. 1 (January 1976): 33–42.

3
Standards for ISDN

The OSI Model

Before looking at the standards that relate to ISDN specifically, it is necessary to look at the general picture of communications systems standards. Communications systems are becoming ever more complex with the emergence of new technologies and the demand for new services (such as video services over fiber optic cables). If such systems are to be compatible, a universal method of defining these systems is needed. The International Standards Organization (ISO) has developed a model for rationalizing the definition of complicated communications systems. This method of fracturing a communications system into its constituent parts is the *open systems interconnect* (OSI) model. Using OSI, numerous communications systems can be broken down, thereby facilitating the definition of standards. This model is the basis for the CCITT X.200 series of recommendations.[1]

The OSI model splits the operation of a communication system into seven layers. By dividing the system into these parts, an understanding of the functions of each part can be gained independently of the system configuration. Each layer defines a different function within the communication system. The layering is organized such that the partitions flow from the interconnection to the electrical, optical interfaces and on to the user interface. Although the ISO model is relatively new, many existing standards fit into this structure. The lower layers of the model are more closely tied to the type of interface used, and the upper layers to the type of service that is performed.

The seven layers of the OSI model are as follows (see Figure 3.1):

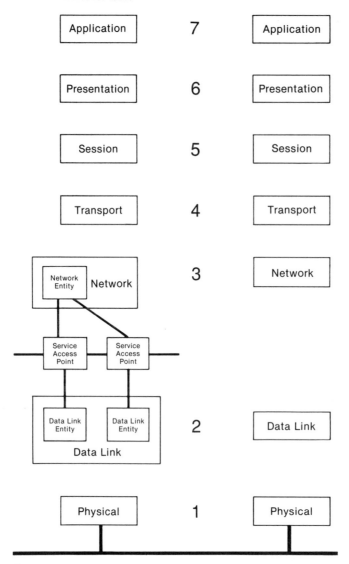

Figure 3.1.   Seven-layer Model.

Layer 1: Physical layer. This layer defines the physical interface that a particular system or network has. The physical layer is responsible for sending and receiving the information across the network.

Layer 2: Data link layer. This layer is responsible for sending and receiving error-free data across the network. Tasks such as error detection and correction are performed by this layer.

Layer 3: Network layer. This layer is responsible for controlling the connections between the various nodes on the network. The establishment of a connection from one node to another is controlled by this layer.

Layer 4: Transport layer. This layer is responsible for controlling the flow of information over the connection. The data that are transmitted are sometimes broken down into smaller blocks at this level.

Layer 5: Session layer. This layer is responsible for controlling the whole operation of connecting to the network.

Layer 6: Presentation layer. The presentation layer defines how the information is presented both to the network for transmission and to the user.

Layer 7: Application layer. The application layer defines the interaction between the user and the communications system.

Each layer can also be divided into elements called *entities*.[2] Entities combine together to form subsystems, which in turn make up the operation of a communication system's functional groups. To perform these functions, the entities from the different layers must be able to communicate. The location in between the layers where the entities converse is called a *service access point*. Because more than one entity can exist within a layer, more than one service access point can be present at an interlayer boundary. The various service access points are delineated by service access point identifiers (SAPI). Entities can only communicate with other entities of the adjacent layer, that is, one above or one below. Entities that are on the same layer are called *peer entities*.

Probably the easiest way to understand how a system can be split into seven layers in this fashion is to look at the following example. In a small grocery store the money is received by the cashiers at the cash register. The money is placed in the register by denomination. Periodically, the money is collected and delivered to the manager's office. At the end of the day, the money is separated and put into bags ready for the pickup from the bank. Normally an armed guard service will be used to collect the money, which is put into special security pouches and then taken by armored truck to the bank. At the bank the security pouches are delivered by the armored truck. The pouches are then emptied, the money separated, and sent to a bank cashier. The cashier will check the money, count it, and then credit the dollar amount to the grocery store's account.

This somewhat simplistic example can be broken down into the OSI model layers in the following way:

Application layer. The money is collected by the cashiers at the grocery store.

Presentation layer. The money is separated into the various denominations in the register. In addition to separating the money, checks and credit card receipts are also segregated.

Session layer. Periodically the money will be collected from the registers and delivered to the manager's office.

Transport layer. The money will be separated into stacks that can easily be handled by the bank. There will also be similar stacks for the checks and credit card payments.

Network layer. The stacks of money are placed into bags containing the address of the bank.

Data link layer. The bags of money are collected by the security service and placed in special security pouches for the bank.

Physical layer. The pouches are taken to the bank in an armored truck.

The process of receiving the money for crediting the grocery store's account can also be split into the seven layers.

Physical layer. The armored truck arrives at the bank.

Data link layer. The security guard delivers the security pouches to the bank.

Network layer. The security pouches are emptied and the bags of money separated. The bags from the grocery store are placed in one stack.

Transport layer. The bags from the grocery store are collected and placed into a package ready for the bank cashier.

Session layer. The bags of money are received by the bank cashier, who empties them. Any packaging material is disposed of.

Presentation layer. The money is sorted, counted, and the total value in dollars is recorded.

Application layer. The cashier logs onto the bank's computer, accesses the grocery store's account, and credits it with the dollar amount recorded.

One point to notice in this example (aside from the factoring into the seven layers) is that there can be more than one entity at various "layers." For example, when the store manager receives the money, it is first sorted out into a stack of cash, then one of checks, and one of credit card receipts. Just as there is a mechanism for sending the cash to the bank there is also one to return the checks and the credit card receipts. The process of splitting the money into different stacks is controlled by different transport entities. The cash is then placed into bags. This is a service access point. So in the above example, several subsystems make up the process of getting the money from the grocery store to the bank, although all the money—cash, checks and credit card receipts—may in fact go to the bank in the same armored truck. This is also true of ISDN, in which the same physical layer is used for different subsystems.

Before leaving the discussion of the OSI model, the topic of *primitives* should be covered.[3] A primitive is an abstract representation of the interaction of the different layers. In the above example, the security guard would tell one of the bank personnel that there is a delivery. This could be a "delivery request indication" primitive from the data link to the network entity. Conversely, once all of the security pouches have been received, the bank employee would sign the relevant paperwork and say good-bye, a "delivery termination request." If the paperwork was in order the guard would respond with a good-bye, a "delivery termination acknowledge." In fact most of the interaction between the different entities across the service access points can be described by using a set of primitives for each layer-to-layer boundary.

To summarize: A communications system can be divided into seven layers as defined by the OSI model vis-à-vis

Application
Presentation
Session
Transport
Network

Data Link

Physical

Each layer can contain one or more entities. Entities of the same layer are called peer entities. Entities interact with each other via service access points. Service access point identifiers (SAPIs) are used to delineate different service access points. Primitives are used to define the interaction between layers and entities.

The various entities function together to form the communications system. These entities, as previously outlined, interact across the layer boundaries using a set of primitives. As described in the analogy, in the sending system, the higher-layer entity initiates an operation. The succeeding lower layers are then stimulated into action. Primitives are passed down through the layers until they reach the physical interface. At the physical interface the process of synchronizing and establishing a dialog between peer entities begins. The physical layers establish a dialog first and then pass a primitive to the data link entity. The data link entity then synchronizes its operation with its peer entity in the receiving system. When the dialog is initiated, a primitive is passed to the network layer. Each layer successively synchronizes and establishes a dialog in turn. When the highest layer

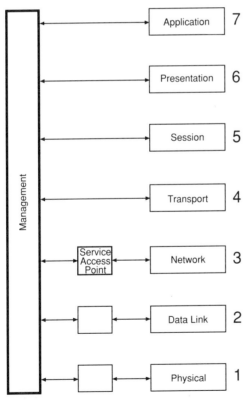

Figure 3.2.   Seven Layers plus Management.

is reached, information can be passed from the transmitting system to the receiving system. In the receiving system, the layers are stimulated from the lowest layer (physical) to highest layer (application). CCITT adds another layer to the OSI model that deals with the *management* functions of the communication system. This layer interfaces to all OSI layers to form a practical implementation of a communications system. This layer would contain, for example, the memory and timer management functions and be similar to an *operating system* (see Figure 3.2).

The various layers are covered by various standards. For example, the Institute of Electrical and Electronics Engineers (IEEE) standard 802.3 covers the physical layer of Ethernet, and X.25 is a CCITT standard for layer 3. In some cases, particularly in the telephone network, these standards vary on a national basis. Although ISDN is moving towards an international set of standards, it difficult to obtain these on government-regulated telephone networks. Consequently, the standards within ISDN that cover the equipment that resides on the end users' property (customer premise equipment, or CPE) are more accepted as international standards. For this reason, these standards will be discussed in more detail. Fortunately, the standards that cover the off-premise equipment are structured in a similar fashion.

CCITT ISDN Standards

ISDN is covered by several different recommendations and standards from CCITT that are specifically pertinent to the lower three layers, that is, the physical, data link, and network layers. In addition to the X.200–X.250 series, which cover the OSI model, the *I series* and *Q series* specifically relate to ISDN. Additional standards will be dealt with that do not specifically apply to ISDN but are used to specify portions of ISDN, for example, voice transmission standards. The X.200 series of standards—the OSI model—has already been discussed. The next series for consideration is the I series.

Reference Configuration for ISDN

Just as it is important to have the OSI model to break down the function of a communications system, it is necessary to break down the various user access points within a network. To do this, the I series uses the *reference configuration*.[4] This configuration is used universally throughout the I series to refer to the different user access points when defining their operation. This reference configuration is shown in Figure 3.3.

The reference configuration shown for the I series concerns the customer premise equipment (CPE). The functional groups shown fit into two categories: *terminal equipment* (TE) and *network termination* (NT), with the term "network" referring to the user network.

Network terminations can have different forms depending on the network that is in place. A domestic telephone network may only consist of one or just a few telephones. A PABX, on the other hand, has many more telephones interfaced to a

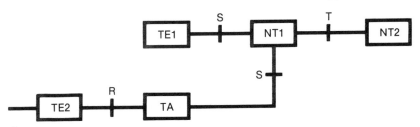

Figure 3.3.   Reference Configuration.

few external lines. The NT is split into two halves to address these differences. NT1 (network termination 1) terminates the network at the transmission line from the off-premise equipment (for example, a CO). The NT1 provides the layer 1 interface between the transmission line, U reference, and the on-premise equipment, T reference point. NT2 (network termination 2) provides elements of layer 1, 2, and 3 functionality depending on the configuration of the user network. When a PABX provides the NT2 function, all three layers are implemented. In the case of simple POTS, NT2 will support only the layer 1 function.

Various types of equipment also exist at the terminal end of the network. Therefore there are two types of terminal equipment and a terminal adapter. Terminal equipment, TE, will perform layer 1, 2 , and 3 functions. TE1, terminal equipment type 1, is a TE that complies with ISDN recommendations. TE2, terminal equipment type 2, is a terminal that meets ISDN recommendations except that it has a different physical interface, for example, RS232. A TA (terminal adapter) provides the interface between the TE2 and the S reference point. Like the NT2, the TA can include one or more of the lower three layers of ISDN to perform the adaption. TAs are sometimes referred to as ISDN modems.

Although the S and T reference points can be similar, there is one major difference between the two. The T interface can only support point-to-point communication. Alternatively, the S interface can support a point-to-multipoint multi-drop architecture, with one NT2 supporting up to eight TEs (see Figure 3.4). Different examples of the types of configurations that can be implemented using ISDN are given in CCITT recommendation I.411 and its accompanying figures; the four most common configurations are given in Figures 2/I.411 (c) and (d) with the physical interface at the S reference point; and Figures 2/I.411 (g) and (h), in which the physical interface coincides with both S and T reference points.

As outlined in the previous chapters, ISDN needs to provide 64 kbs^{-1} channels at various reference points to support both voice and data services. These chan-

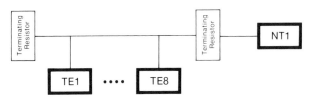

Figure 3.4.   Point to Multipoint Configuration.

nels are referred to as *B channels*. To facilitate the signaling on the ISDN physical interfaces a separate channel is provided, the *D channel*. There are two types of access allowed on ISDNs depending on the number of B channels that are supported. *Basic access* provides two B channels and a D channel, commonly referred to as 2B+D. This type of access is sometimes referred to as the *basic rate interface* (BRI). The D channel has a bit rate of 16 kbs^{-1}. *Primary access* can have either 23 B channels plus a 64 kbs^{-1} D channel (in the United States), or 30 B channels plus one 64 kbs^{-1} D channel (in Europe). This type of access is sometimes referred to as the *primary rate ISDN interface* (PRI). The types of accesses allowed are covered more fully in recommendation I.412.

I.430: The Physical Layer for Basic Access

The ISDN network is connected by several different physical interfaces, which are specified in a similar manner. Rather than cover each one in detail, one will be chosen as an example. The specifications for the other interfaces will be mentioned to highlight the salient points. The interface chosen for detailed analysis is the S interface for basic rate access because it has been more fully defined due to its international acceptance. The specification that covers this interface is recommendation I.430, and consists of different parts, that is, electrical, operational, and testing. This specification will be outlined in these categories.

Electrical Specifications

This interface is intended to operate between the TE and NT. In practical terms this means operation between the end user's equipment and the PABX or external telephone line interface. Consequently, the cabling that is used is the existing telephone line, that is, the four wires commonly found in North American telephone installations. This cabling can be used because the loss in signal over a typical line length is acceptable for ISDN purposes: 6 dB over a 1 km cable.

One quality that is rare about this interface is its ability to operate in point-to-multipoint configurations. One NT can be used to terminate up to eight TEs on a single S interface connection. This is realized practically by having a multidrop architecture with each TE connecting across the telephone line. There are different specifications that pertain to the different configurations.

To provide the BRI, the S interface must allow 2B+D access of 2 × 64 kbs^{-1} + 16 kbs^{-1} giving a total of 144 kbs^{-1}. In reality, a higher bit rate on the line is needed to perform additional functionality. The S interface operates on a four-wire interface, with two wires used for transmission from NT to TE, and two for TE to NT and clock information obtained from the data. The S interface uses *baseband transmission*; only one transmission frequency is used, so the 2B+D channels must be time-multiplexed on the line. To enable the time multiplexing, a framing signal must be extracted from the incoming data. Other bits are needed to provide maintenance functions between the two layer 1 entities. The final outcome of these individual constraints is a bit rate on the line of 192 kbs^{-1}.

The S Interface Frame Format

The 192 bits are made up into frames of information that are sent over the S interface. In many of the other ISDN physical layer interfaces the data on the line are framed. Each frame of information contains 48 bits. These are as follows (see Figure 3.5):

In the NT-to-TE Direction

Bit	Symbol	Function
1, 2	F, L	Framing signal with balance bit
3–10	B1	Information bits for B channel 1, 1st octet
11	E	Echoed TE D channel bit
12	D	D channel bit to TE
13	A	Activation bit
14	F_A	Auxiliary framing bit
15	N	N bit
16–23	B2	Information bits for B channel 2, 2d octet
24	E	2ND E channel bit
25	D	2ND D channel bit
26	M	S1 bit
27–34	B1	B1, 2d octet
35	E	3RD E channel bit
36	D	3RD D channel bit
37	S	S2 bit
38–45	B2	B2, 2d octet
46	E	4TH E channel bit
47	D	4TH D channel bit
48	L	Frame balance bit

In the TE-to-NT Direction

Bit	Symbol	Function
1, 2	F, L	Framing signal with balance bit
3–11	B1, L	B1, 1ST octet with balance bit
12, 13	D, L	D channel bit with balance bit
14, 15	F_A, L	Auxiliary framing bit with balance bit
16–24	B2, L	B2, 1ST octet with balance bit
25, 26	D, L	2ND D channel bit with balance bit
27–35	B1, L	B1, 2ND octet with balance bit
36, 37	D, L	3RD D channel bit with balance bit
38–46	B2, L	B2, 2ND octet with balance bit
47,48	D, L	4TH D channel bit with balance bit

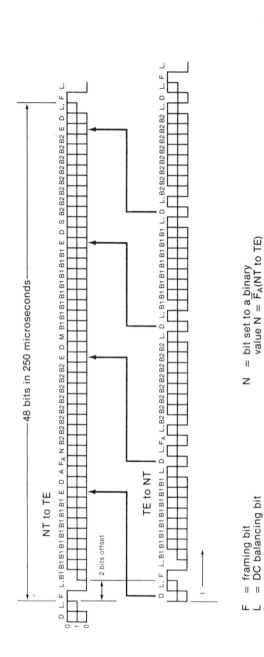

F = framing bit
L = DC balancing bit
D = D channel bit
E = D echo channel bit
F_A = auxiliary framing bit

N = bit set to a binary
     value N = $\overline{F}_A$(NT to TE)
B1 = bit within B channel 1
B2 = bit within B channel 2
A = bit used for activation
S = S channel bit
M = multiframing bit

Note: Dots demarcate those parts of the frame that are independently DC-balanced. The F_A bit in the direction TE to NT is used as a Q bit in every fifth frame if the Q channel capability is applied. The nominal 2-bit offset is as seen from the TE. The corresponding offset at the NT may be greater due to delay in the interface cable and varies by configuration.

Figure 3.5. Frame Structure at Reference Points S and T.

At first glance the arrangement of the frame needed to transfer the voice/data information over the B channels is complex. However, as the function of each of the groups of bits is described in more detail, the operation of the S interface will become more apparent.

The first step to understanding the function of some of the bits is to describe the data that are actually transmitted on the telephone line. The line code chosen for this purpose by the CCITT I.430 standard is *pseudoternary coding*. This code is a variation of the alternate mark inversion (AMI) described in Chapter 2. The logic 0 is represented as a positive (+ve) or negative (−ve) pulse on the line and logic 1 is represented by a zero level on the line. Hence the term "pseudo," as one of the levels is zero. To maintain the DC balance on the line to ensure good transmission, the +ve and −ve pulses are alternated. When a stream of binary data is encoded in this fashion and transmitted on the line the overall DC level on the line will be zero.

If for some reason a +ve pulse (or high mark) is followed after one or more zeros (or spaces) by another high mark, then there is a violation of the line code. Under normal circumstances this would be an error in the transmission; however, in the case of the S interface, code violation errors are deliberately introduced. The deliberate errors delineate the boundaries of the 48-bit frames. To ensure that the violation that is detected at the receiving end is a framing mark and not a randomly induced error, a second violation is introduced in the frame. This second violation can occur either due to the B or D channel information or is introduced in the auxiliary framing bit.

The framing bit will be set to a logic 0 and can be transmitted as a high or low mark. Similarly, the framing balance bit is a logic 0 and in this description will be a low mark. For this description a high mark will be assumed. In the case of NT-to-TE, the first logic 0 in the B, D, or E is transmitted as a low mark. A code violation will be detected—two sequential low marks. The A bit normally would be a logic 1. In the case in which there are no logic 0s in the B, D, or E channels, the code violation is introduced with the auxiliary framing bit. This F_A bit ensures that a pair of code violations will occur within less than or equal to 14-bit intervals. The same is also true for the TE-to-NT direction.

One of the difficulties to overcome in this type of framing by code violation is maintaining the DC balance on the line. To solve this balance bits are used to even out the transmitted pulses. In the case of the NT-to-TE frame only one bit is used for the frame. In the case of the TE-to-NT frame, individual framing bits are used for the different channels. In addition to DC balancing the frame, the balance bits also determine that the last logic 0 that was transmitted is of the same sense, +ve or −ve, as the framing bit of the next frame. Thus the code violation of the start of the next frame is assured.

The feature of point-to-multipoint operation on the S interface can cause problems for the transmission scheme. One of the first consequences is the necessity for individual channel balance bits in the TE-to-NT direction. Suppose one TE were connected very close to the NT and a second TE connected at the far end of the line. Now let the first TE be transmitting on the B1 channel and the second on the B2 channel. The signal levels of the B1 and B2 channel information will be different. So each of these channels must be balanced individually. In the case of the other bit in the frame, the marks will be superimposed and again must be individually balanced.

Another difficulty of the point-to-multipoint configuration is the handling of the D channel. Because in the previous example there are two B channels allocated to two different TEs, the signaling will have to go to both TEs. With two transmitters trying to transmit on the same D channel, errors can result due to collisions. To solve this, a collision resolution scheme is used. When a logic 1 is transmitted, a transmitter has a high input impedance to the line, and the transmission of a mark on the line by another transmitter will override it. Therefore, if one TE is transmitting a space (logic 1), and the second a mark (logic 0), the mark will be received at the NT. To acknowledge that the mark has "won," the NT transmits the received D channel bit from the TE on the E channel. By examining the received E channel bits, each TE can tell if a collision on the line has resulted in one TE getting the line (see Figure 3.6). The use of the activation bit will be covered in the section on activation.

Finally, the N and S bits need to be described. The F_A bit, in addition to being used to generate the extra code violation, is also used for multiframing. In this process a special pattern is sent over the F_A bit to allow the S bits to be used. If the frames are numbered with reference to a repeated pattern on the F_A bit, then information can be sent repetitively at a slow speed using the S bits in more than 2-bit groups. The N bit is then used to balance the pattern on the F_A bit. The type of information that is transferred in this fashion is maintenance data between the two physical link entities in the NT and TE. These bits are used in the T1D1 Standard[5] in the following manner (see Figure 3.7).

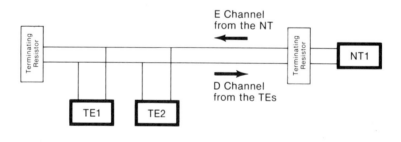

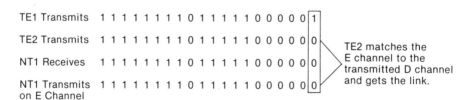

Figure 3.6.   Collision Resolution.

S-Channel Structure

Frame Number	NT-to-TE F_A bit Position	NT-to-TE M Bit	NT-to-TE S Bit
1	ONE	ONE	SC11
2	ZERO	ZERO	SC21
3	ZERO	ZERO	SC31
4	ZERO	ZERO	SC41
5	ZERO	ZERO	SC51
6	ONE	ZERO	SC12
7	ZERO	ZERO	SC22
8	ZERO	ZERO	SC32
9	ZERO	ZERO	SC42
10	ZERO	ZERO	SC52
11	ONE	ZERO	SC13
12	ZERO	ZERO	SC23
13	ZERO	ZERO	SC33
14	ZERO	ZERO	SC43
15	ZERO	ZERO	SC53
16	ONE	ZERO	SC14
17	ZERO	ZERO	SC24
18	ZERO	ZERO	SC34
19	ZERO	ZERO	SC44
20	ZERO	ZERO	SC54
1	ONE	ONE	SC11
2 etc.	ZERO	ZERO	SC21

Figure 3.7. S-Channel Structure.

Activation

Because there are only two wires in each direction of the S interface, it is necessary to extract timing information from the incoming information stream. To initialize the extraction process, a *synchronization procedure* is used during the activation sequence. To achieve synchronization in both directions, four types of signal are defined for the S interface (see Figure 3.8):

Info 0 No signal
Info 1 TE to NT only: Pattern of +ve mark, —ve mark, and 6 spaces
Info 2 NT to TE only: Frame with B, D, and E bits = logic 0; activation (A) bit = logic 0; balance bits set to balance frame
Info 3 TE to NT only: Normal synchronized information on the line
Info 4 NT to TE only: Normal synchronized information on the line; activation (A) bit = logic 1

The info 1 and info 2 are used to initiate the activation sequence on the line. If the line is deactivated (info 0 in both directions), then the TE will start the activation procedure by sending info 1. In the case of the NT, info 2 will be sent. The station at the receiving end has many options as to how to proceed depending on the events leading up to the reception of the info signal. To keep a record of the preceding events, the state of the NT or TE must be known at a given point

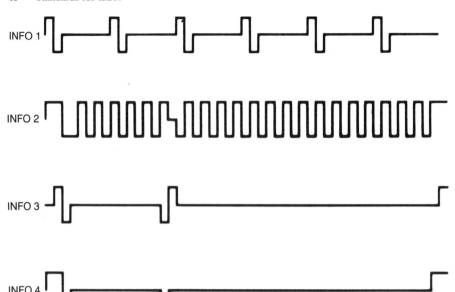

Figure 3.8.   INFO Signals.

in time. External events will be recorded as the state of the NT or TE changes. Various stimuli, for example, the reception of an info signal, will cause the state to change. The states that can exist in the TE are called *F states*, and in the NT, *G states*.

The F states for the TE are described in I.430 as follows:

F1 No power
F2 Power on but input signal on the line not determined
F3 Info 0 being received; TE deactivated
F4 Info 1 transmitted to start activation; awaiting response from NT
F5 Signal received after sending info 1; signal type not yet resolved; transmission of info 1 terminated
F6 Info 2 received; info 3 transmitted; waiting for info 4 from NT
F7 Activated; info 3 transmitting and info 4 received
F8 Frame synchronization lost

The G states for the NT are as follows.

G1 Deactivated; sending and receiving info 0
G2 Awaiting activation; info 2 is transmitted
G3 Activated; info 4 is transmitted and info 3 received
G4 Deactivation sequence started awaiting info 0 from TE (only the NT can initiate a deactivation sequence; however, the line can be deactivated after certain errors occur.)

In the above description of states, it can be seen that the TE or NT can activate the line. A set of primitives is used to start activation. The line activation is requested by a higher data link layer by invoking a *physical activation request* (PH-AR). After this has been issued, the activation sequence will commence. If successfully completed, a *physical activation indication* (PH-AI) is passed to a layer 2 entity. To deactivate the line a *physical deactivation request* (PH-DR) is given from layer 2. This can only be invoked in the NT. When deactivation is completed, a *physical deactivation indication* (PH-DI) is issued to layer 2 (see Figure 3.9).

In addition to the OSI seven layers, CCITT defines a *management layer* that fits onto the side of the seven layers. Primitives can be transferred between the management entity and the physical layer. These control the operation of the equipment when an error is detected. A *management-to-physical error indication* (MPH-EI) is issued by layer 1 if a framing error occurs or is recovered. A *management-to-physical error response* (MPH-ER) will cause the layer 1 to abandon its attempt to attain framing synchronization.

To ensure the successful operation of the interface, a set of timers is defined in this physical layer. The timers handle the problem of an activation sequence that is started by an NT with no TEs attached to the line. After sending the info 2 signal for a defined period of time, the timer will time out and a PH-DI will be issued to layer 2.

Together with the info signals, the states, and the primitives, the operation of the physical layer can be generally defined. This definition is done in two ways in I.430. One is in a tabular format, the *finite state matrix*, which cross-references various states with different stimuli that can occur. The next state that is reached after a given stimulus can be read off this table. Alternatively, a flow chart representation is available. This is written in *functional specification* and *description*

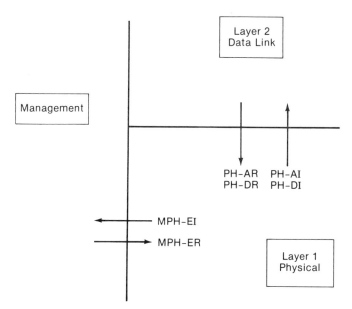

Figure 3.9.   Layer 1 Primitives.

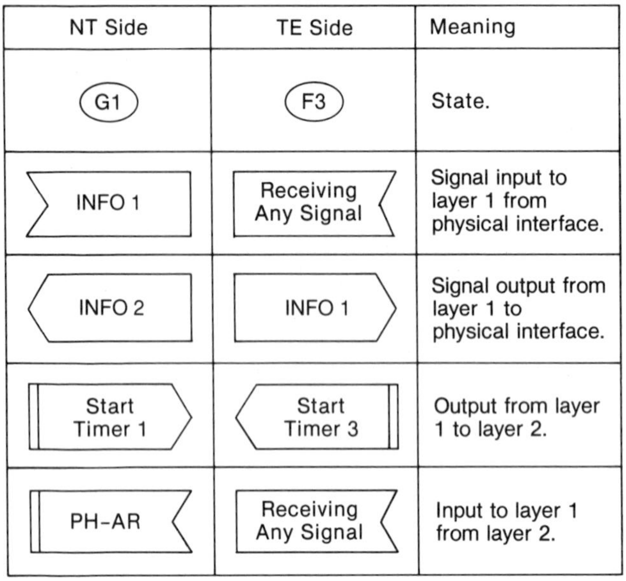

NT Side	TE Side	Meaning
(G1)	(F3)	State.
INFO 1	Receiving Any Signal	Signal input to layer 1 from physical interface.
INFO 2	INFO 1	Signal output from layer 1 to physical interface.
Start Timer 1	Start Timer 3	Output from layer 1 to layer 2.
PH–AR	Receiving Any Signal	Input to layer 1 from layer 2.

Figure 3.10a.   SDL Symbols.

language (SDL) that is defined by CCITT. An example of this type of flow chart is given in Figures 3.10a and 3.10b.

As with any form of transmission, it is important to ensure good transmitter and receiver design. This true for both analog and digital systems. The pulses on the S interface use *pseudoternary code modulation*. To specify these pulses in the conventional sense would require a table of parametric values. Instead, a template is used to define the pulse shape that the system must output. The template can cover several design parameters. For example, the rise time for the pulse is important and could either be defined as a voltage change per unit of time or as a region inside a template. By defining the shape of the pulse, the receiving eye diagram can also be defined as the characteristics of the telephone line are known. By using the eye diagrams for the different points on the transmission line, the best sample point and the range of the receiver inputs can be determined. Here is a list of parameters that the pulse template defines for the S interface:

Maximum and minimum pulse amplitude
Maximum and minimum rise and fall times
Maximum and minimum pulse width and frequency
Maximum pulse overshoot on the rise and fall of the pulse

There are two pulse templates defined in the I.430 specification: one for a 50 Ω load and one for a 400 Ω load. The first pertains to a point-to-point configuration in which there is zero length of transmission line. The two 100 Ω termination impedances are taken together in parallel to give the 50 Ω loading. The second loading value is used to simulate the configuration in which eight TEs are on the

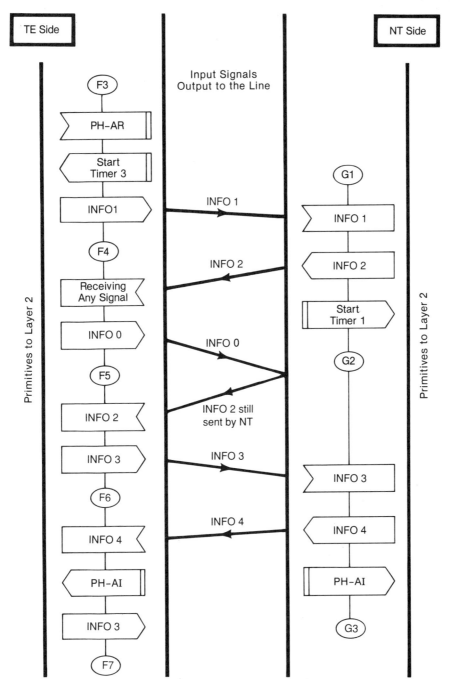

Figure 3.10b.   Activation Example.

same S interface. Only the 50 Ω template is shown in Figure 3.11, together with the areas that define the various parameters.

Another template included in the I.430 specification is the impedance template. The purpose of this template is to provide a specification that covers a number of parameters. There are in fact three impedance templates. One is for the input impedance of the receiver/transmitter of the TE. The second is for the NT transmitter. Both assume the transmitter is in the high impedance state, that is, transmitting a logic 1. The input impedance is defined across a frequency spectrum. This assures that the output pulse from the transmitter is not unduly effected by other receivers or transmitters on the line. If the loading on the line is below these values, the eye patterns on the line would be affected. This would cause an incorrect pulse shape to be received, which would alter the bit error rate of transmission and impair the quality of information carried across the line. The third template is for the case of the TE/NT transmitting a logic 0. This is set to a value of > 20 Ω. Several examples of the expected eye diagrams are given in the I.430 specification.

Many references have been made to the point-to-multipoint configuration. Some special specifications relate solely to this configuration. There are two types of multipoint configuration: *short passive bus* and *extended passive bus*. The

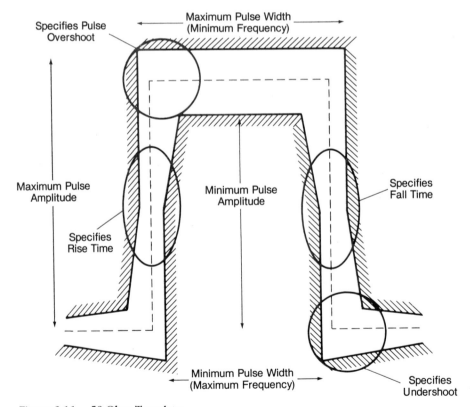

Figure 3.11.   50-Ohm Template.

major difference between the two is the distance limitations between the NT and TEs. In the short passive bus, the distance is chosen such that the pulses from the TEs will superimpose upon each other, ensuring minimum destructive inter-ference between TEs. In the extended passive bus, the distance between the TEs is restricted. The maximum distance constraint ensures minimum interference between the transmitters of TEs on the bus. The round-trip delay of the system defines the maximum values for the point-to-multipoint configurations. As the delay is a function of the transmission line impedance, different lengths will result from different types of transmission lines. The line lengths and the round-trip delay calculations are given in annex A of I.430.

In all configurations, the TE derives clocking information from an incoming pulse stream transmitted by the NT. The TE output is clocked with this extracted timing information. Because the clock is derived from the incoming bit stream, jitter is added to the output signal. The jitter specification is split into two param-eters: jitter due to clock extraction, and jitter on the output signal that causes phase deviation between the input and output signals. In the first case a figure of $\pm$ 7% of a bit period is given, and is measured using a high-pass filter with a 30 Hz cutoff. The $\pm$ 7% refers to the maximum amplitude of the jitter signal, with the second stipulation being a measure of the jitter signal's frequency content. In the second case, a jitter specification is given with respect to the input signal. In this case however, the figure is only a definition of the jitter amplitude. The jitter specifications of the input test signal for this measurement are also given both in terms of the amplitude and the frequency content.

One other jitter specification is given for the NT. Even though the NT sources the system clock to the S interface, this clock is often derived from another system clock. For example, for an NT that is realized as a line card in a PABX, the S interface will be phase locked to the PCM highway system clock to ensure syn-chronization of the bit streams between the systems. Jitter can be induced in this phase locking step. For this reason, NT jitter is limited to $\pm$ 5% of a bit period with a high-pass filter of 50 Hz. This correlates with the specification of the input signal for the TE phase deviation measurement.

The S interface has the capability of supplying power to the terminal equip-ment in one of three ways: the power can be supplied locally (for example for a voice/data workstation based around a PC); it can be supplied from the NT along a separate pair of wires; or it can be supplied across the four-wire S interface. This is done by using a *phantom feed technique*. One side of the DC power is fed in a common mode fashion along one pair of the wires and the other side down the second pair. In this way, DC power is supplied along the four wires in addition to information transmitted digitally. The transmitter and receiver are coupled to the physical interface using an *isolation transformer*. If the line side of this trans-former is center tapped, then the DC power can be supplied using the two center taps (one for receive, and one for transmit).

By supplying power in this way, a four-wire interface can be used for a simple telephone service. Both the digital information and the power can be supplied over the same four wires. However, one parameter is very important when this type of power supply is used. The center tap of the line winding in the interface transformer must be located in the center of the winding. If the center tap is off

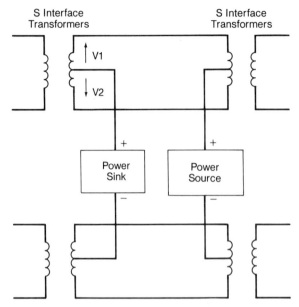

Figure 3.12.   Longitudinal conversion loss.

center, then the current flow through the two halves of the winding will not be matched (see Figure 3.12).

This will cause a DC offset in the transformer and affect the performance of the digital transmission. The measurement of the accuracy of the center tap of the transformer is called the *longitudinal balance*. A specification for this parameter, together with suggested test environments, is given in I.430.

The Data Link Layer

The layer that resides above the physical layer is the data link layer; this layer is responsible for packaging information from higher layers. The packaged information is passed to the physical layer for transfer. In an ISDN application, there are typically two types of subsystem present: call control (which must now be done digitally) and data transfer. Although the physical layer concerns itself with both 2B+D channels, this is not the case for layer 2. In a voice/data terminal there will be at least two layer 2 entities. Depending on the type of data transport mechanism that is chosen, these two entities can be quite similar or radically different. Fortunately, the call control entity is standardized.

The S interface, as with any of the ISDN interfaces, must support existing standards and applications, especially the existing data standards. In current analog transmission, a seven-layer structure will exist for data transfer. The only difference for ISDN transmission is the physical interface. This will alter only layer 1 and its interaction with the other six layers. In a call control application, the system is unique to the telephone networks. Although much has been copied from data applications, there are many metamorphoses to produce a system suited to call control. A standard has been internationally defined for the call control layer 2, which is based on the X.25 data network.

Layer 2 for call control is more commonly called the *link access protocol D channel* (LAPD). The two standards that cover this specification are I.440 and Q.921. I.440 describes the interface in more general terms, whereas Q.921 provides more specific details. The LAPD provides the interconnection between the network layer (layer 3) and the physical layer. This interface is done via the service access points. In the physical layer there are physical service access points. The data link service access points interface to the network layer. The interface between the physical layer and the LAPD has already been discussed from the layer 1 perspective. In a similar way, the interface to the network layer is described using various primitives. The LAPD is used to provide the following functions to the NT or TE:

- Handles more than one data link connection—to control two calls across one S interface
- Frame alignment, and packaging of upper-layer information into the correct format
- Sequence control of multiframe transmission/reception of long data messages
- Detection of transmission errors across the interface
- Error recovery from errors where possible and notification to the management entity when an unrecoverable error is detected
- Flow control of data frames being passed between peer data link entities

LAPD uses high-level data link control (HDLC) protocol to packetize and transfer data. Data from layer 2 or the higher layers are built up into information frames and transmitted on the D channel.

Each frame has an opening flag, address field, control field, optional information field, CRC, and a closing flag. To maintain integrity of the delimiting flag (0111 1110), a logic 0 is injected into the data stream when a pattern of six logic 1s is detected in the frame. In the discussion of the frames it is assumed the data under discussion have not had zero insertion performed (see Figure 3.13).

The first part of the frame is the address field, which is broken down into two octets. The first octet is composed of the service access point identifier (SAPI), the command/response (C/R) bit and the extended address (EA) bit. The EA bit is used to signify whether the octet is the last in the address field; in LAPD there is always a second address byte, so this bit is set to logic 0. The C/R bit is used to signify whether the frame is a command or a response. From the user side, TE, the C/R bit is set to a logic 0 for a command and a 1 for a response. At the network side a logic 1 signifies a command and a 0 a response.

The SAPI can be used to define one of several service access points depending on the frame destination. The ones most commonly used for LAPD are as follows:

   0: For call control procedures
   1: For user data on the D channel
  16: For packet control procedures
  63: For management procedures

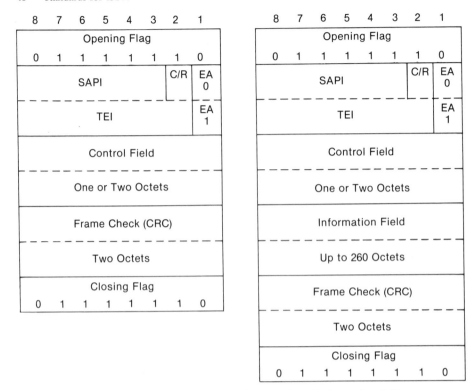

Note: Frame formats are shown with and without optional information fields.

Figure 3.13.   LAPD Frame Formats.

SAPI 1 is used when data are multiplexed onto the D channel when the D channel is not being used for call control purposes. For example, because the 16 kbs^{-1} is not continuously required for the call procedures, slow speed data, 9,600 baud say, can be multiplexed onto the D channel. The SAPI determines to which subsystem the packet belongs (Figure 3.14).

The second octet of the address field contains a seven-bit TEI and an EA bit. The EA bit in this case will always be set to logic 1 to signify that this is the last octet of the address field. A TEI can have a value between 0 and 127. However, the TEI value 127 is reserved for broadcast information and cannot be used as a unique identifier.

The next field in a LAPD frame is the control field. This field can contain one or two octets depending whether modulo 8 or *modulo 128* operation is selected. In certain cases, layer 3 information is split into several layer 2 frames. When this is done, each frame is numbered to maintain the sequence of the multiple frames. For modulo 8, the sequence numbers can have values 0 through 7. In modulo 128, these numbers range from 0 to 127. The control field length will vary from 1 to 2 octets to accommodate the different sequence count values.

There are three different types of frame that are defined by the control field, supervisory (S), information (I), and unnumbered (U) frames, which perform different functions in the layer 2 entity. S frames are used to supervise the frame

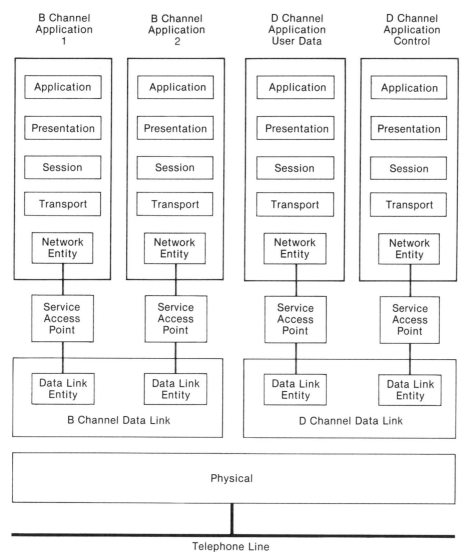

Note: Frames for different services multiplexed on the physical interface and demultiplexed using different SAPI values D and B channel data are separated by time division multiplexing.

Figure 3.14.   Use of Multiple Protocol Links on the S Interface.

transfers. I frames are used to transfer layer 3 information across the interface. U frames are used to provide additional data link functions.

The control field also contains the *poll/final* (P/F) bit. The P/F bit has two functions. In a command frame, a P/F bit is used as a poll to solicit a response from a peer entity. In a response frame, a P/F bit is used to signify that a response has been transmitted in answer to a poll request. In both cases, the P/F bit is set to a logic 1 to stimulate the poll/final condition.

In LAPD protocol, frames on the D channel can be sent in either *acknowl-*

edged or *unacknowledged modes*. In an acknowledged transfer, frames received are acknowledged by a special S frame sent from the receiving end that contains the sequence number of the next expected frame in the sequence. If the received frame is out of sequence, a reject S frame is sent to the sending data link entity. This method of acknowledging frame reception is called *asynchronous balance mode* for modulo 8 and *asynchronous balance mode extended* for modulo 128. A U frame is used to set this mode of operation, that is, *set asynchronous balance mode* (extended) or *SABM(E)*.

If the balance mode is set, then each frame is acknowledged. In some cases this can be a disadvantage. Consider a satellite link in which the round-trip delay is on the order of quarter of a second. Each time a frame is sent, it will take more than 250 μs before it is acknowledged, thus severely reducing the effective transfer data rate. To compensate, acknowledgment is only sent after a certain number of frames have been received. For instance, seven frames could be transferred before an acknowledgment is sent. The number of frames that can be received before an acknowledgment is sent is called the window size. In this example, the window size would be seven. Therefore, because every frame is acknowledged, LAPD has a window size of one.

Layer 2 is initiated by a request from layer 3, instigated by layer 3 issuing a primitive, *DL-ESTABLISH*. The data link entity will have to check which state it is in, for example, has the physical link been activated? When DL-ESTABLISH is issued for the first time, the physical link will not be active and a PH-AR is issued to the layer 1 entity. After receiving a PH-AI from layer 1, layer 2 can change the state of the *physical link control* to reflect the change in line status. The next action required is to assign a TEI value to the TE. TEI values are assigned by the NT, which keeps a record of these values to ensure that two TEs do not have the same TEI value.

Because a TE does not initially have a TEI value, the broadcast TEI (127) is used until one is assigned. The management SAPI is used in the first octet of the address field to signify that the TEI assignment is a management function. The TE will first "suggest" a TEI value, normally generated from a random number generator. If this value has not been already allocated, the NT grants the value to the TE. Otherwise, a reject is given and the TE must try a different TEI value. Once allocated the TE must store this value and use it in future frame interactions (see Figure 3.15).

One final component to layer 2 is a set of timers. It is not the responsibility of layer 1 to inform layer 2 of an error on the transmission line. The notification is given to the management entity. Instead, layer 2 sets timers to time the response from the peer entity. For example, in a TEI assignment procedure, a timer is set when a TEI value is requested from the NT. If no response is obtained, then the timer times out and the correct action is taken (in fact, the TEI request is retransmitted). To supplement the timers, a set of counters is required for the operation of layer 2. These are either counters for sequence number counting or retransmission counters to monitor the number of times frames are retransmitted. These timers and counters can either be realized in software or by specific pieces of hardware.

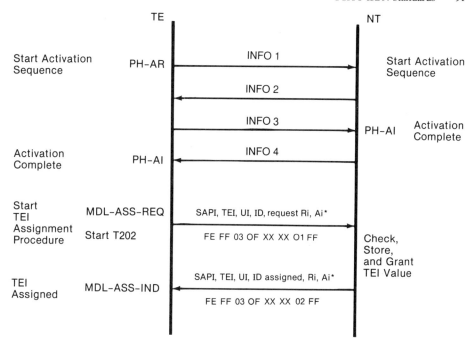

Note: D Channel frames mnemonics at the top, hex at bottom (no flags or CRC shown).

Figure 3.15. TEI Assignment.

The operation of the Layer 2 entity will be dealt with in more detail in the succeeding chapters.

The layer above the data link layer is the network layer. This layer is responsible for establishing the connection between the TE and NT. This layer provides the following functions:

Allocation of the B channel
Alerting the called party (ringing)
Sending dialing information
Activating the data link layer
Checking the support of a requested service

The network layer is used to establish and control the telephone call across the network. In the case of a simple telephone, this layer will control nearly all the features provided to the user. In NT equipment, layer 3 will handle control of the interface between the line and the PCM highway. As with layer 2, operation of this layer will be dealt with in the succeeding chapters.

Voice Digitization: G.711 and G.712

The practice of converting voice to a stream of digital information is not new and has been in common usage since the mid-1960s. CCITT has developed a series of standards—G.711 and G.712—to define the measurement and performance of such networks. As with the S interface, these standards provide templates that stipulate performance limits of a digitization process, covering various elements of analog/digital systems such as compression algorithms; frequency performance; gain performance; and noise levels. The first point, compression algorithms, is very important because of its effect upon the standard. The compression algorithm dictates not only the type of compression used but also the type of units and loading used for various parameters.

As outlined in Chapter 2, a compression algorithm is used to give a good dynamic range with an 8-bit code. The best algorithm is a logarithmic one. However, because the function $\log x$ diverges for small values of x, the compression must become linear for small values of x. The actual compression used is a modified logarithmic algorithm. Two such modified algorithms are used. In North America and Japan the compression algorithm is μ-*law*, in Europe *A-law* is used. In addition to both algorithms being modified to take into account the small signal level values, the algorithms are also a *piecewise approximation* to the logarithmic scale. The scale is split into sections, or *chords*, and a linear approximation is derived.

One important fact about encoding: in many cases the values used for standard signal levels are not what might be expected. For example, a digital value of all zeros will not translate into a signal level of 0 V. This is important when considering test patterns or the effect of loading digital values into a voice channel. Additionally, the B channels on the S interface are set to FF when not in use. If this is fed into a codec, unexpected results can occur. One test pattern often used in voice signals is the *digital milliwatt* (mW). This is a pattern that will result in a power of 1 mW output from the codec (see Figure 3.16).

Another effect to be considered in the digitizing process is the performance of the codec itself. The codec will have its own transfer function. These performance parameters could affect the voice signal as it is digitized. To minimize the effect of the codec transfer function, G.712 specifies the coder/decoder performance. Several parameters are covered in this specification including the frequency, phase, and gain performance over ranges used by a voice channel. As with the pulse specifications, these parameters are given as a template. The figures are given for only half duplex transmission, that is, one path for transmission and one for reception with the performance measured by connecting the codec PCM output to the codec PCM input in two stations. Therefore, for a complete system two measurements must be taken, one for each path through the two stations.

In many cases, the template is normalized to a given frequency of 1,000 Hz. The practical value used for normalization may not be the 1,000 Hz specified. By chosing a value that is slightly different, the effect of interference due to the test signal being a submultiple of the PCM highway clock is reduced. In most cases, the performance of a codec will be given with respect to G.712 templates. As more than one test method is stipulated in some of the templates, the testing method should also be given when stipulating the template and performance results.

As pointed out in Chapter 1, the range of human speech varies from person to

Table of Values
Loaded into Codec

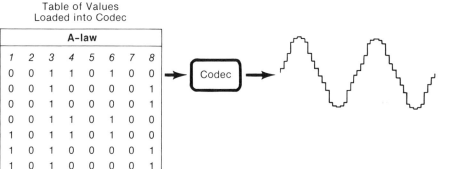

A-law							
1	2	3	4	5	6	7	8
0	0	1	1	0	1	0	0
0	0	1	0	0	0	0	1
0	0	1	0	0	0	0	1
0	0	1	1	0	1	0	0
1	0	1	1	0	1	0	0
1	0	1	0	0	0	0	1
1	0	1	0	0	0	0	1
1	0	1	1	0	1	0	0

μ-law							
1	2	3	4	5	6	7	8
0	0	0	1	1	1	1	0
0	0	0	0	1	0	1	1
0	0	0	0	1	0	1	1
0	0	0	1	1	1	1	0
1	0	0	1	1	1	1	0
1	0	0	0	1	0	1	1
1	0	0	0	1	0	1	1
1	0	0	1	1	1	1	0

Figure 3.16.   The Digital Milliwatt.

person. Codecs must operate over a large dynamic range. The gain through the system must be constant across this range. It is important to measure the relative gain in a system for different input levels. There are two methods outlined in the G.712 specification for the measurement of the gain versus input level.

G.712 specifies two types of noise: *idle channel noise* and *signal-to-total distortion*. Idle channel noise is specified as −65 dBm0p. The unit that is used signifies that the measurement is made with respect to 1 mW at the 0 transmission level using *psophometric* loading. Another unit of noise measurement often used is dBrnc0 (pronounced "debrenco"). This unit measures noise with respect to a reference noise signal, virtual silence, of 1 picowatt (pW). The loading used is called *C message weighting*. For conversion purposes, −90 dBm is equal to 0 dBrn. The two different types of loading have been developed as a result of the different compression methods. Each load is intended to simulate the characteristics of the human ear.

The signal-to-total-distortion specification is a template that defines the minimum performance allowed. This specification covers not only noise and distortion due to codec analog sections but also quantization noise added in the analog/digital conversion.

Data Transfers on the ISDN Network

One of the functions of ISDN is to integrate the transfer of voice data. The G.711 and G.712 specifications cover digitized voice; however, the specification of data networks is not so straightforward. Data have been transferred across the telephone network using many different methods outside of ISDN. For example, analog modems can handle transmission speeds of 9,600 baud or greater. Single 56 kbs^{-1} or 64 kbs^{-1} channels in a T1 link are used for data transfers. ISDN has to address these types of data transfer in addition to any new standards generated by the facilities offered by ISDN technology.

Before delving into the different methods of transferring data, one important quality of data transfer should be understood. There are two types of data routing in the telephone network, including ISDN. One method is *circuit-switched data*; the second is *packet-switched data*. In circuit-switched data, a call is handled in the same manner as a voice call. The sending station dials up the receiving station. Once a connection is established through the network, a circuit is made such that data are transparently passed from one station to the other. In certain cases, particularly ISDN, the terminal's speed underutilizes the bandwidth of the communication media. To enhance the utilization, packet switching is used.

In this type of switching, data are sent using a packet protocol—for example, LAPD. At the transmitting end, more than one station is connected to the circuit connection. Each station is given its own address, with each transmitted packet containing the address of the terminal. The receiving end routes the data packets depending on the embedded address. By using this type of transmission scheme, more than one terminal can be connected by using a single connection through the network. This technique is also useful if several lines are connected in a star configuration—terminals connected to a mainframe for example.

Packet switching also gets around another phenomenon introduced by ISDN. In the analog telephone networks, data speeds are limited to rates of around 9,600 bits per second. With packet switching, several terminals can use single 64 kbs^{-1} channels offered by ISDN. The packet switch gives a form of *rate adaption* from 9,600 bits per second to 64 kbs^{-1}. Each terminal connected to the B channel gets its turn in sending a packet of data to its corresponding receiving station. There are, however, two disadvantages to this type of data handling: the requirement for a relatively large amount of software to control the "virtual" link through the network, and a buffer to store packets waiting for a gap in the channel. When the terminal is not busy, a special sequence is output to allow other terminals to gain access to the link. In many cases this is either an idle channel signal (all ones) or a continuous sequence of flags. These are known as *idle channel stuffing* or a *flag stuffing* rate adaption, respectively.

A second method available for rate adaption from the slower terminals connected to ISDN networks simply adds extra bits in the data stream to pad out the data rate. Additional data are also added to the bit stream to allow end-to-end signaling and flow control. This method is known as a *bit stuffing rate adaption* and has the advantage that significantly less software and memory are needed.

The disadvantage is that only one terminal can be used on the B channel, resulting in a waste of most of the 64 kbs^{-1} bandwidth.

The most common standard used for packet switching is X.25/X.21, which gained a great deal of acceptance even before the coming of ISDN. In fact, the call control procedures used on the D channel for setting up a telephone call are very similar to X.25 procedures. One of the standards used for bit stuffing protocols is V.110,[6] which outlines different bit stuffing procedures to support various standard terminal rates. One standard emerging as a general data transfer standard is V.120. This standard covers several types of rate adaption schemes that will allow many different types of *data terminal equipment* (DTEs) to be accommodated by ISDN.

One other method involves a hybrid approach of an analog modem connected to a codec. The digitized analog data are transmitted through the network, allowing modems to get good performance through the network with minimum impact upon the installed base of equipment. There would be no difference between the quality offered by an ISDN voice connection used for data and one used for voice. In the analog network, special analog lines are sometimes used that have better-than-average performance to transfer modem data. The analog channel provided by ISDN should offer a measurable performance increase even over the conditioned data lines used in analog networks.

Even though ISDN offers many advantages in terms of point-to-point performance for data transfers, there are also other advantages that can be exploited. In addition to using the D channel for call control, it can be used to transfer low-speed packet data, in the range of 2,400–9,600 bs, in addition to the 128 kbs^{-1} offered by the two B channels. Data call and flow control can also be performed over the D channel and can increase the effective data rate on the B channels. Other "data" services can be offered, for example, access to a computerized telephone book. This would make the telephone book look like a data base. So even the voice services can be augmented by the data capability of the ISDN.

References

1. *CCITT Red Book*, vol. VIII, fascicle VIII.5, recommendation X.200, section 6.
2. *CCITT Red Book*, vol. VIII, fascicle VIII.5, recommendation X.200, section 5.2.1.3.
3. *CCITT Red Book*, vol. VIII, fascicle VIII.5, recommendation X.210, section 3.2.3.
4. *CCITT Red Book*, vol. III, fascicle III.5, recommendation I.210, figure 1/I.210.
5. American National Standards Institute (ANSI) Integrated Systems Digital Network. Basic Access Interface for S and T Reference Points: Layer 1 Specification.
6. CCITT recommendation V.110.

4

The ISDN Terminal

Functional Parts of the Terminal

The design and function of subscriber equipment will undergo key changes due to ISDN. ISDN offers an opportunity to construct voice/data workstations more easily. The ISDN terminal has different functional blocks depending on the services supported. In general these blocks are the line interface; the voice interface; a microprocessor system for control and signaling; and data interface circuitry.

To integrate these functions into a single piece of equipment, it is important to first choose a good system architecture. Many device suppliers have defined interchip architectures to allow easy connection of ICs. Generally these interface specifications revolve around a serial bus structure with time division multiplexed channels similar to PCM. The channels in the bus have two types of functions—one set for data and one for control information—and are designed for use in network as well as terminal equipment. By optimizing the bussing structure, the amount of microprocessor interaction can be minimized. One of the most common busses is the *general communication interface* (GCI). This interface, based on the Siemens *ISDN Orientated Modular* (IOM®) architecture, is supported by several companies. To date Siemens, Plessey, Alcatel, Italtel, and Advanced Micro Devices are either producing or developing devices to interface into this architecture.

The GCI is designed to interface several different functional blocks of a piece of ISDN equipment. Channels are created in the bus structure to allow the devices to communicate with each other. The bus is based on a channel containing four octets. Two octets are for the B channels, one octet for monitor and contention resolution, and one octet for the D channel and layer 1 control. The bus is built up using varying numbers of channels, to a maximum of eight.

The Siemens IOM® bus can be used in either a point-to-point mode or a point-to-multipoint mode. In the point-to-point mode, there are four octets to provide connectivity between the two devices. Two octets contain the B channel data, one contains a monitor channel, and one holds the control channel. The control channel contains the D channel bit from the ISDN interface, plus a control nibble (4 bits long) to transfer control to, and indications from, the line interface. The monitor channel is used to carry the monitor bits and the D channel collision status from the line interface.

In point-to-multipoint mode, several of the four octet structures are built up to provide connectivity between devices. A maximum of eight of these structures can be concatenated to form a 2,048 kbs bus. In the point-to-multipoint mode, the monitor and control channels can be used for additional functions. The monitor channel can transfer interchip information (for example, initialization data). The control channel can transfer device status information. In both point-to-point and point-to-multipoint modes, the monitor channel controls the contention on the bus when trying to access the various channels (see Figures 4.1a and 4.1b).

The bus structure from MITEL, the ST-Bus™, contains 32 octets per frame on a time division multiplexed highway. The highways can be used for either control information or data. In the case of control highways, the information transferred is dependent upon the device at the receiving end. The bussing can be used for point-to-multipoint operation.

Northern Telecom and Motorola's IDL can be used in either point-to-point or point-to-multipoint architectures. The frequency of operation depends on the type of structure used. In point-to-point operation there are two B channels, one

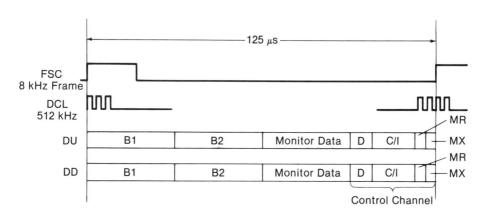

B1, B2: Circuit switched voice/data

D: D channel for signalling and packet switched data

C/I: Command/indication bits for control

MR, MX: Monitor channel control bits

Note: DCL is programmable for values from 512 kHz to 8.192 MHz (DU and DD shown for DCL = 512 kHz).

Figure 4.1a.   Frame Structure of the IOM Rev. 2 Interface.

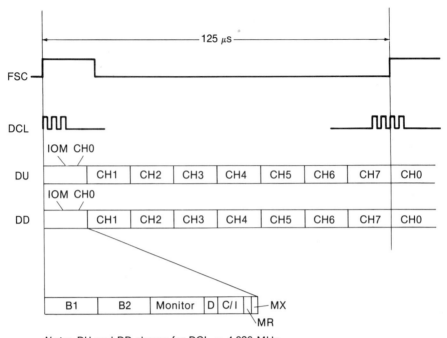

Note: DU and DD shown for DCL = 4.096 MHz.

Figure 4.1b.   Multiplexed Frame Structure of the IOM Rev. 2 Interface.

D channel, one M channel, and one A channel. These channels are used to transfer data between devices. The B and D channels are analogous to the ISDN 2B+D; the M is for maintenance purposes, and the A is for an auxiliary bandwidth. The B channels are 64 kbs, the D is 16 kbs, and the M and A, 8 kbs each. This gives a total bandwidth of 160 kbs.

In the point-to-multipoint mode of operation, the channels are built up as slave devices are added. A maximum of 16 channels can be allocated in this way. This gives a total bandwidth of 2,560 kbs required for a maximum of 16 slaves.

All of the various bussing structures use a frame period of 125 μs. This ensures that digitized voice information can be carried by the bus and permits easy integration with devices designed for digital exchanges. Many devices suitable for ISDN have been available for some time. These components were originally designed for digital exchanges. However, they can also be used in the digital telephone needed for ISDN; this is particularly important when designing circuits to interface between existing subscriber equipment and ISDN. The interface is essentially moved from the exchange to the terminal adapter.

The ISDN Line Interface

Before ISDN became the standard for integrated services, PABX vendors started to evolve proprietary schemes for voice/data terminals. However, most of these solutions consisted of analog techniques to integrate the two types of information. ISDN is a fully digital interface and so the line circuitry is new even to the designer of "digital" telephones. Although there are subtle differences between the different line interfaces, there is a lot of commonalty in the design. A line interface, be it S or U, will require the following components:

Line interface circuit;
Transformer(s);
Protection circuitry; and
Connection to the line (connector, termination, etc.).

The line interface circuit is sometimes referred to as the *line interface unit* (LIU). In many devices the LIU is part of the complete IC. This is particularly true for the S interface, in which the LIU is less complicated compared to the U interface.

Depending on the reference point, the LIU will have a different structure due to the additional technical problems that have to be overcome to transmit full duplex digital information over a two-wire line. There are two approaches to this problem. One is a technique called *time compression multiplexing* (TCM); the other is *echo cancellation*. In TCM, after both stations are synchronized, the information is transmitted in one direction first, and then, after a settling time delay, in the other direction. In this way the line is "turned around" after each transmission. Only one station will be transmitting at one time. In echo cancellation, both stations are allowed to transmit simultaneously. The receiver constructs a replica of the echo from the transmitter and subtracts it from the received signal. In theory only the signal from the sending station will remain.

The advantage of TCM is simpler receiver design. The receiver does not have to handle the transmitter echo. However, to be able to transfer the same amount of information, the baud rate on the transmission line is considerably higher, more than twice the bit rate due to the settling time. This higher frequency of transmission reduces the transmission length that can be achieved. Also, in North America there exists the phenomenon of *bridge taps*, a spur connected in parallel to the telephone line that is used for party line services. A bridge tap will add echoes to the received signal, thus making it difficult to design a TCM receiver that will give adequate performance. Echo cancellation will give a lower baud rate and consequently a longer line length, adaption to bridge taps, and full duplex transmission. The cost of this improved performance is a substantially more complex receiver design.

The Line Interface Circuit

No matter which reference point is considered, the most radically changed part of a telephone design will be the line interface. Several components are required to implement this interface. Although the main component will be the line interface unit itself, the external parts used will play an important role in the performance of the complete circuit and hence in the overall performance of terminal equipment. Unfortunately, the CCITT recommendations and other line interface standards pertain to terminal equipment and not the individual components. It is important to look at the line interface in detail and to understand the role played by the active and passive devices.

The main considerations in an interface design are basically the same for different interfaces:

Compliance with the pertinent line interface standard;
Clock extraction circuitry;
Interface to the layer 2 devices; and
Layout considerations for low-noise operation.

To begin with, a line interface design using the Siemens *PEB2080 S interface bus circuit* (SBC) will be studied. The PEB2080 contains the major portion of the circuitry needed to implement an S interface layer 1 (see Figure 4.2). The device handles interfacing between an S reference point and the Siemens IOM®. The LIU will perform the translation of signals from the line into 2B+D data and control information for the line. By referring to Figure 4.2, it can be seen that the part has five functional blocks: the receiver, the transmitter, the clock extraction circuit, the IOM interface, and the activation and line control. This type of structure is used in most line interface devices.

The receiver and transmitter sections will vary quite substantially across the

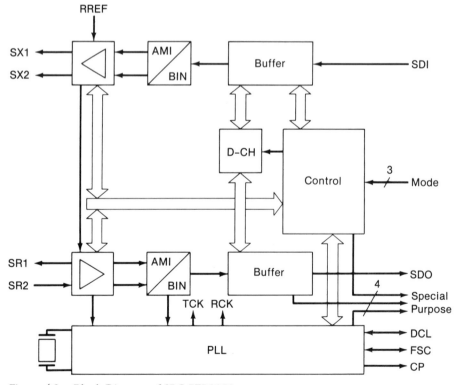

Figure 4.2.   Block Diagram of SBC PEB2080.

different reference points. In the case of the TE S interface reference, there are some unique requirements for the transmitter/receiver. They must both be able to handle a point-to-multipoint configuration.

For the transmitter this involves two specifications: meeting the 400 Ω template, and meeting the input impedance template of CCITT recommendation I.430. The 400 Ω template is used to simulate the load seen by the TE transmitter if seven other TE transmitters were on the line together with the NT receiver and the terminating resistors. By careful design of the transmitter circuitry, it is possible to meet this template. The second requirement is a more stringent design problem.

In a situation in which a terminal is connected to an S interface in a point-to-multipoint configuration, its transmitter must present a minimum impedance to other transmitters, ensuring that their pulse shapes are maintained. This minimum impedance value must be maintained even when the power is removed, or the TE has to be unplugged from the line when powered off. If a complementary metal-oxide semiconductor (CMOS) transmitter circuit is connected to the line with the power turned off, then pulses from the other transmitters can be "shorted out" by the powered down transmitter. One way to solve this is to add a relay between the line and the transmitter. When the power is removed, the relay de-energizes and disconnects the transmitter from the line. Alternatively, this problem can be overcome by specialized transmitter design. The PEB2080 has such a transmitter design and will not load down the line when connected to a point-to-multipoint configuration without power to the device.

The receiver has equally unique design challenges when used in point-to-multipoint operation. In this configuration, several TEs can be connected to the line. A contention algorithm is used by comparing the E channel bits to the transmitted D channel bits. The receiver must be able to perform this task and output a signal if a mismatch is detected.

The receiver is also responsible for "squaring up" the input signal. The line input is first passed through a filter, and is then sampled. The input sampling is performed adaptively, that is, the threshold point is a percentage of peak input level. If this percentage is too high, then there is less noise immunity for the receiver. If the percentage is too small, then the threshold can be close to the noise floor when receiving signals at a maximum line length. A zero crossing detector is used to give a reference to the start of a bit period. The sample point is set to a fixed delay from a zero crossing. For an S interface this is about 80% of a bit period (4.2 μs).

Once an input signal has been reconstituted back to a TTL level, the clock extraction can be performed. The clock extraction circuit performs two functions: bit clock extraction and frame clock extraction. A high-frequency *digital phase-locked loop* (DPLL) is used for this purpose. By using a crystal oscillator circuit, the DPLL is able to meet the jitter tolerance requirements for both the receiver and transmitter; frequencies of several megahertz are used (7.69 MHz for the PEB2080). When an external crystal is used for this purpose, it is important that the crystal be decoupled with two identical capacitors. This will ensure correct start-up of the oscillator and will maintain the frequency tolerance. If these capacitors are omitted, the LIU will have difficulty extracting the clock; the DPLL will not function correctly, resulting in loss of synchronization both at the TE and NT.

Both the receiver and transmitter connect to the line via a transformer that provides isolation from any DC levels on the line. Nonetheless, it does have a contributory effect upon LIU performance as a whole. In the transmit direction, the transformer can affect the final pulse shape in many different ways. For example, if there is too much distributed capacitance in the transformer, the rise and fall times of the pulse will be increased. These could deteriorate to an extent in which the pulse would not meet the specified pulse templates. If the winding ratio is incorrect, an incorrect pulse level can be delivered to the line. When the transmitter goes from transmitting a mark, high or low, to a high impedance space, the stored charge in the transformer and associated circuitry must be dissipated. If there is too much leakage inductance and stray capacitance in the transformer, then this dissipation process can cause an unacceptable amount of overshoot and violate the pulse template (see Figures 4.3a and 4.3b).

It is important that the correct transformer selection be made for a given line interface device. This will ensure proper operation to the transmitter and receiver. In many cases suppliers of LIUs will be able to supply a list of recommended transformers in addition to a list of recommended crystals.

In many instances a DC offset voltage is used to level shift the received line signal. This enables the receiver to operate differentially while only using a single rail supply. When examining the pulse shape at the chip side of the transformer, this must be borne in mind. Additionally, the transmitters can operate in a differential mode. This will result in unusual waveforms if an oscilloscope is used to measure the voltage between one of the transmitter outputs of the IC and ground. If an oscilloscope is used to monitor an interface, then care must be taken to ensure that it does not unduly load down the line. This can be done accidentally by creating a ground loop when connecting the scope probes. The easiest way to overcome this is to connect the scope to the line via a transformer (see Figure 4.4).

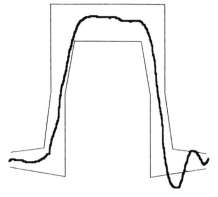

Figure 4.3a.   Pulse Overshoot due to
Too High a Value of Stray Capacitance.

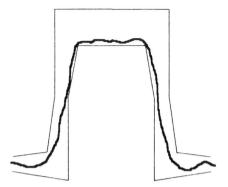

Figure 4.3b. Low Pulse Amplitude due
to Bad Transformer Ratio or Low Output
Driver Voltage.

Another part of the line interface is the *protection circuit*. This circuit dissipates any surges on the line due to high-voltage sources such as lightning. These circuits are needed for both S and U interfaces. The surges involved vary from specification to specification but fall into three types: a common mode surge, differential mode surge, and a constant current surge. In all cases the protection circuit should dissipate the power of the surge, and prevent any harmful voltages reaching the user or destroying the TE circuitry. The usual method of implementing a protection circuit is to use a zener diode in a back-to-back configuration across the line. Care must be taken when selecting the diodes that the protection circuit does not cause an interface to violate the impedance standard. Examples of protection circuits are given in Figures 4.5a and 4.5b.

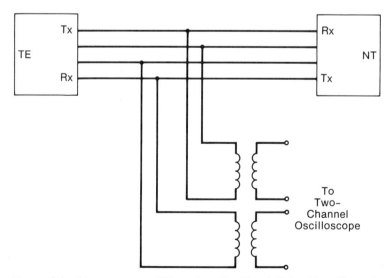

Figure 4.4.   Measurement of S Interface Line Signals Using Two-Channel
Oscilloscope and Isolation Transformers.

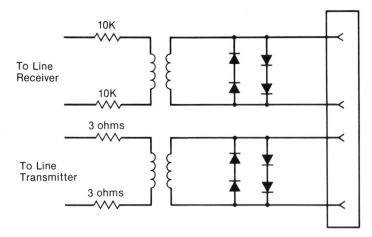

Figure 4.5a.   S Interface Protection Circuit.

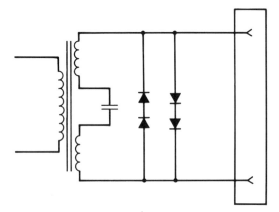

Figure 4.5b.   U Interface Protection Circuit.

Finally, the interface circuit must connect to the line. For the S interface this is done using an eight-, six-, or four-pin modular connector. The connection diagram given in the CCITT specification refers to generic pole positions for the connector. The pole positions are then cross-referenced to the pin numbers for different connectors. A diagram of a connector is given in Figure 4.6 together with the cross-reference list for S interface connections. For the S interface, care must be taken when deciding the position of the termination resistor. A common practice is to include this resistor as part of the TE design; yet many TEs are intended to operate in point-to-multipoint configurations. To allow this, a switch must be included such that the resistor can be switched out on a terminal used in a multipoint configuration. Only the terminal furthest away from the NT should have a termination resistor in circuit. Because the U interface connector is controlled by the governing bodies of the various countries, it will vary from country to country.

A complete circuit for an S interface is given in Figure 4.7. This circuit contains the LIU, the crystal, the transformers, the protection circuit, and the connector.

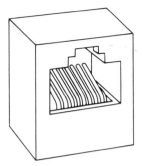

Pin No.	TE	NT
1		
2		
3	Transmit	Receive
4	Receive	Transmit
5	Receive	Transmit
6	Transmit	Receive
7		
8		

Figure 4.6.   S Interface Connector.

Siemens also manufactures a transceiver for the U interface. The functions of this device and the associated circuitry are similar to the S interface device. The 2B1Q line-coded signals are translated to TTL-level signals on the IOM bus. The control of the interface is accomplished by using the control channel of the IOM. The major difference between the applications is the complex design of the echo cancellation circuit and filter.

Many semiconductor and system manufacturers are working on solutions for the U interface. These devices translate the line signals from the U interface to the bus standard of the manufacturer. The task that faces the designers of these devices is twofold: they must design an adaptive echo cancellation circuit to work at the data rate required for ISDN, and also an analog front end to convert

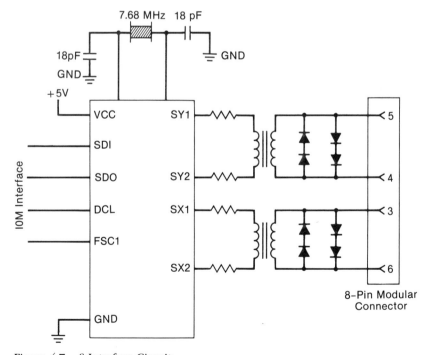

Figure 4.7.   S Interface Circuit.

incoming line signals to a digital word for digital signal processing. These two steps will need a major amount of research and development. The amount of investment in both time and money may serve to slow down the deployment of ISDN on the U interface. However, as will be seen in chapter 8, there are other solutions to utilizing ISDN services off-premise.

The Telephone Interface

ISDN involves digitization of voice information in subscriber equipment. This moves the analog/digital conversion from a line card back to a telephone. The treatment of signaling is another important change in the telephone functions of subscribers' equipment. Because signaling is handled in the D channel, the telephone must now provide functions that it never had to before—for example, ringing. In current analog telephones, ringing is supplied by a line card as a high-voltage AC signal. All a telephone has to do is switch this signal, using a diode bridge, to an electromechanical bell. For an ISDN telephone, the AC signal is no longer provided by the exchange. Instead, a local oscillator must provide the ringing signal. The ringing is initiated by a message transferred from the exchange to the terminal equipment over the D channel.

Although this is a trivial example, as many phones now use electronic ringers, it does serve to show the point that functionality is being moved from the responsibility of the exchange to the telephone. Another example would be the dial tone. Currently the dial tone is exclusively supplied by an exchange. In ISDN, the exchange no longer has to provide this function and in certain cases might not even be practically able to do so. Consider the case of a point-to-multipoint S interface with three TEs connected. Suppose that TE1 and TE2 have already successfully established a call over the telephone line with TE1 being allocated B1 and TE2, B2. If TE3 suddenly goes off-hook, then a B channel cannot be assigned. It is impossible for the exchange to signal this condition to the subscriber by a tone as there is no B channel to transfer it. So without local tone generation, the telephone would appear "dead."

There are, broadly speaking, two types of approach taken to providing telephone interfaces in ISDN terminals. One school of thought holds that ISDN should use new equipment, including the telephone. The other is that existing telephone equipment should be integrated into ISDN networks by using a telephone adapter. The two different applications require different approaches. In an ISDN telephone the circuitry is much simpler. However, the telephone must be designed and qualified. This may be a challenge that designers new to telephony might not want to undertake. Where an ISDN telephone is implemented, care must be taken in all aspects of the analog design.

There are two different types of codec available on the market today. One is a *switched capacitor filter codec* and the other a *digital signal processing codec* (DSP). In the first type a capacitor is alternatively charged and discharged to realize the filter function. In the second, a *finite impulse response filter* (FIR) is used to realize the filter function. In both codec structures, devices have been primarily designed for exchange line card applications. Nevertheless, they are finding a new lease on life as the digitization point is moved from the line card to the telephone. The biggest advantage of DSP codecs is the ability to match different types of telephone networks throughout the world without a component change. This feature is extremely important when designing a telephone adapter for several different national telephone networks.

Additionally, new codecs are being designed to meet the requirements of the ISDN telephone. For example, the main audio processor of the AMD Am79C30 and the Siemens PSB2160 contains not only codecs and filters but also loudspeaker drivers, ringers, and tone generators. These are particularly useful features when designing an ISDN telephone.

A key feature that should be reviewed is *hook-switch monitoring*. In a standard telephone this is performed by a line card in the exchange. In an ISDN telephone, irrespective of which type of implementation is chosen, this function must be performed by a microprocessor. This may not sound too difficult until the problem of debounce is considered. A considerable amount of software overhead may be utilized by implementing this function in hardware. The AMD Am79C30 offers a solution to this by providing a debounced hook-switch input that will give a microprocessor interrupt when the hook switch has changed state for a required period of time.

One other feature added to the new ISDN codecs is DTMF generation. This may seem a useless function because dialing information can be provided over the D channel of an interface. But for interfacing to existing equipment, a DTMF generator is essential. Selecting a long-distance carrier is one example, or activating a home telephone message recorder so that messages can be accessed remotely. In both cases the telephone dialing into the exchange must be able to provide DTMF tones.

Finally, *voice mail* can be easily added to ISDN telephones. This is particularly true if the ISDN telephone service is part of a voice/data workstation application or plug-in card. It is important that a codec used in this application be chosen with an interface that can be easily integrated into the microprocessor architecture. A byte has to be transferred every 125 μs, so it is desirable that this interface circuitry have direct memory access (DMA) capability.

Layout Considerations

The layout of ISDN circuitry brings with it its own associated problems. The main difficulty in realizing ISDN circuits is the opposing challenges of mixing analog and digital circuitry. On the one hand, the analog interface to the telephone cannot have a high-noise component in its output signal. Conversely, the digital line interface and microprocessor system use high-frequency clock signals. The layout of these circuits must take the interaction of the two subsystems into consideration.

Although different types of terminals may be developed, there are certain design rules that can be laid down to ensure that a high-performance analog design can coexist with a high-speed microprocessor.

Perhaps the most important factor in the design of a mixed digital/analog

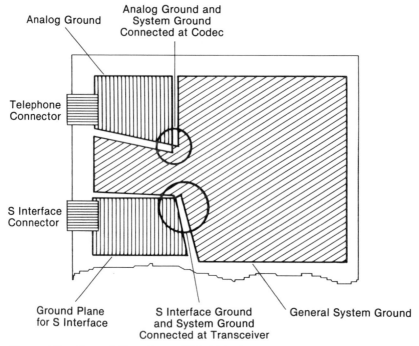

Figure 4.8.   Ground Plane for an ISDN Telephone.

system is the power and ground layout. The use of power and ground planes in a multilayer fabrication is desirable. This reduces the difficulties of power supply decoupling. One additional technique that can help to reduce the cross-coupling of noise from these various systems is the use of isolated ground planes within the ground plane layer. The various ground planes can be joined together at a single point. This can be a star point at the power supply or a ground pin of one of the components (see Figure 4.8).

Positioning of the components themselves can also influence overall system performance. For example, high-frequency signals should be kept away from analog input lines. This may sound easy enough but consider the case of a plug-in card for a PC. The telephone connector must be placed close to the S interface connector and the PC backplane connection. It may be difficult to avoid running an analog line close to an S or U interface output.

Noise can be picked up from other unexpected sources. The receiver for the line interface could be susceptible to high noise levels from digital clocks. Care must be taken to make sure that no stray clock signals are superimposed on a digital line input. This noise may not stop the line interface from functioning, but could have a considerable effect upon the bit error rate of the interface.

The Function of the Microprocessor

The inclusion of a microprocessor is essential to any ISDN design. Many feature telephones now include microprocessors, but the amount of processing power required for ISDN is far in excess of the needs of a standard feature telephone. The processor must perform several important functions in the ISDN telephone:

Control of layer 2 and 3 protocol software
Formatting HDLC data
Control of telephone functions (off-hook, dialing, etc.)
Control of data services
Control of line interface
Self-tests

Each one of these tasks requires a program to implement it. There are many operations that require an intensive amount of processing. This can be realized by either a microprocessor program or as a function of an ISDN device.

Processing of layer 2 and 3 data is a case in point. A great deal of processing is needed to implement these functions.

For example, the amount of code for this could be in excess of 48 kbytes of program for a simple telephone application. This requires that a major portion of the processing capability be allocated to this function. The layer 2 and 3 functions can be separated into their individual processes. (This will be dealt with in Chapter 7.) However, the requirements for layers 2 and 3 can influence the hardware design.

The use of timers is common to layers 1, 2, and 3. Timers are used throughout the operation of call control software for many purposes. In layer 1, for example, a timer is used to discern if the other end of the interface has responded in time. In certain cases, more than one timer can be in operation. Because the implementation of timers is an arduous task for the software designer, any aid that can be offered to solve timer implementations is greatly appreciated. These timers can be implemented in may forms. Many devices that contain HDLC functions contain a timer (AT&T UNITE chip, Siemens PEB2070 ICC, etc.). These timers can be used for layer 2 functions such as a timeout for reception of an acknowledgment from the peer entity after a frame has been sent, or as a *watchdog timer*. In the latter case some devices do offer a timer that acts as a reset control. Many line interface devices will perform some of the timer function required by I.430. This is done autonomously by the device without any microprocessor intervention. Other timer functions can be performed by timers on the microprocessor system itself. Many microprocessors and microcomputers include a timer block as part of the chip design. Unfortunately, in many cases these timers are already allocated for operating system functions.

Another function required by layer 1, 2, and 3 is *counting*. As with the timer function, counting can take different forms. In certain cases, counters are included as part of an ISDN device. Two types of counting can be performed by these devices. One type is in the counting of bits and frames on the line interface for synchronization purposes. In all cases this is included in a line transceiver. At the S interface, for instance, the device must detect three out-of-synch frames before an error is issued. The second type of counting relates to the more traditional type of application. In both layer 2 and 3 a count must be kept of the number of times a frame of information is retransmitted. If this number exceeds a certain value, then an error is generated. Certain HDLC controllers contain this counter. In this case the counter could be implemented in software; however, it is much more convenient if this is done in hardware.

In the previous two examples, frame retransmission was mentioned. Memory is required to store frames both for normal and retransmission purposes. Received frames must also be stored in some type of memory. In data transfer applications, large blocks of memory may be required for temporary storage of information while the transmission or reception process is in progress. One type of memory that is becoming increasingly more common in communications systems is *dual port memory*. These memories are used to store data between a communications controller and a host processing system. For instance in a PC ISDN card, data from the PC can be transferred in short bursts at high speeds, as in the case of a floppy disk interface. The communications interface must operate at the same speed as the line, so data are stored in blocks in a dual port memory.

A similar solution that rivals the dual port memory architecture is *direct memory access* (DMA). This offers access to larger areas of memory by a communications device. One popular device that uses this type of approach is the Intel 8530 serial communications controller. As well as offering timer functions, many microprocessors and microcomputers also contain a DMA controller as an integral part of the device.

Even for a POTS implementation of an ISDN telephone, a relatively large amount of memory is needed. Care must also be taken in which type of memory is used. *CMOS static RAM* can be costly, although as more advances are made the cost should come down. *Dynamic RAM* is cheaper but requires a controller and consequently more power. In certain instances, for example, the storage of a TEI value, the memory must be able to operate without power.

Another use for memory is program storage. Even though many line interface standards are now defined, this is not the situation for layer 3 (and to a certain extent layer 2) call control software. One solution is to determine which type of exchange the TE is to be connected to, then download the layer 2 and 3 protocol software. When considering the amount of memory for a given application, the amount of RAM needed for program space must also be added into the equation.

The control and formatting of HDLC frames require a great deal of processing. These functions, if performed by a microprocessor, can take a substantial portion of the available processing power. Fortunately there are many HDLC controllers available on the market that can provide varying levels of support, depending upon the application. All of these controllers will provide the following functions:

Zero insertion and deletion
Flag insertion, deletion, and detection
CRC generation and checking
Transmission and reception abort detection/recognition
Interframe fill

With the exception of the final point these are covered in chapter 2. *Interframe fill* is the character that is transmitted between frames. This can be of one of two forms: all 1s or a series of flags. In some devices this is programmable; in others it is not. In certain cases the CRC algorithm is programmable; this is the case with the Intel 8530.

Other advanced features, some of which are specific to ISDN, are as follows:

Recognition of management and programmable SAPIs
Recognition of management and programmable TEIs
Processing of control field
Frame sequence numbering
Handling of nonunary window sizes
Collision detection for multiple access

These functions can be handled by a microprocessor. In more advanced terminals it is more advantageous to have these functions handled by a dedicated piece of silicon. A data terminal that can send and receive data and control data transfers from a dumb terminal must be able to handle the call control and data protocol for 2B+D on an ISDN link. This involves a considerable amount of processing. Any of the more mundane functions, such as the ones already listed, that can be performed by the ISDN device will allow either the choice of a less powerful, and consequently cheaper, microprocessor, or the implementation of additional user features.

Different types of memory are required within an ISDN terminal design. This can be realized in many forms, for example, static or dynamic RAM. In many microprocessor architectures it is possible to move memory in blocks far more efficiently than by a byte at a time. For a communications system the data are transmitted onto the line a byte at a time. To cope with these two stringent requirements HDLC controllers include a *first in first out* (FIFO) memory. These memories vary in size depending on the HDLC controller.

For data to be transmitted, information is first moved from a microprocessor to a FIFO using a *block move instruction* (the implementation of this in reality will vary from processor to processor). Once data are in the FIFO, the HDLC controller can transmit it on the line. In the situation in which a frame of data is longer than the FIFO size, an interrupt must be issued by the FIFO when it is ready to accept more data (see Figure 4.9). To stop a memory location being written to at the same time as the HDLC reads information out, *contention logic* is included as part of the FIFO.

In the receive direction, information from the line is processed by an HDLC controller and then written into the FIFO. When the controller is ready to allow a microprocessor access to the data, an interrupt is issued. The microprocessor can then perform a block move of data into its own memory (see Figure 4.10).

The size of the FIFO structure, the type of frame received/transmitted, and the speed of the incoming data determine how fast the microprocessor must react to interrupts. The FIFO size determines how many bytes can be processed by the HDLC controller before either a phantom data byte is transmitted or a received byte lost. The type of data and line speed determine how fast data are transferred to and from the FIFO.

Another method of transferring data to and from HDLC controllers is the use of DMA. This method has the advantage that large amounts of information can be transferred without microprocessor interaction. The disadvantage is that a specialized piece of hardware is needed to implement DMA control. In certain cases, the DMA controller will be part of the microprocessor. The number of DMA

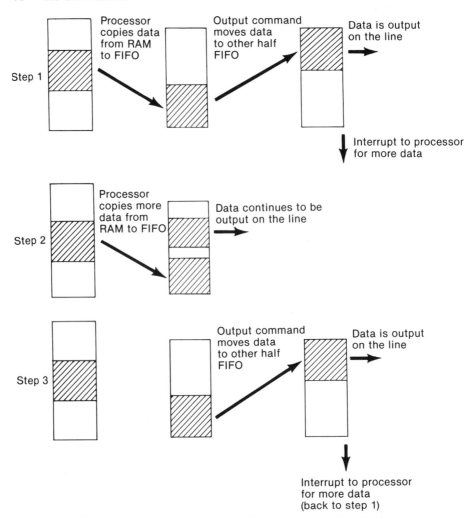

Figure 4.9.   Transmission of an HLDC Frame Using Siemens HLDC Controllers.

channels required will depend on the application. For instance in a voice/data workstation, two DMA channels will be necessary: one for the D channel and one for the B channel. These could be supported by a 80188-type processor. When two data channels are used, then three DMA channels are required. An additional DMA controller chip is needed to handle this number of channels.

One disadvantage of DMA data transfer is that the controller setup may become burdensome for small packets. In call control protocol software, the average packet size may be less than 32 bytes. This would mean that for a packet of 32 bytes or less, say 5 bytes, several register operations would be required to set up a DMA controller. This may take more time than just writing the bytes to a FIFO. This will be dealt with more fully in the next section on data flows.

It can be seen that the choice of the microprocessor is influenced by several

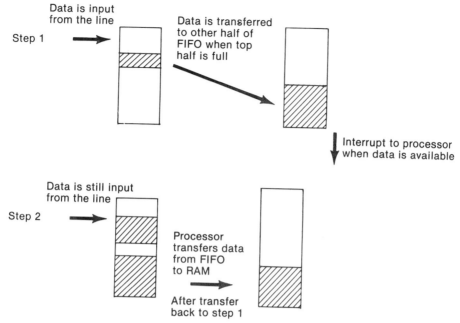

Figure 4.10.   Reception of an HLDC Frame Using Siemens HLDC Controllers.

factors. The type of application can determine the processing power essential to realize a complete design. The functions provided by peripheral devices also determine the microprocessor power. The type of data flow structure used for the application DMA or FIFO will determine whether a microprocessor with peripheral components on board is advantageous. Finally, the power supply limitations can have an influence on the choice of microprocessor. This will influence not only the type but also the technology that is used in the device fabrication. In low-power applications, such as a priority telephone, it would be mandatory to use a microprocessor manufactured with CMOS technology.

Data Flow within an ISDN Terminal

ISDN allows the addition of data services to the digitized voice information. The speed at which data can be transferred on an ISDN link is far in excess of that available using a modem but considerably less than a local area network (LAN). The design of the data flow through a terminal is extremely important to fully utilize the microprocessor potential. A bad data flow would lead to the selection of a microprocessor with too much computing power.

In modem applications, data speeds of 9,600 baud are becoming quite common. With the addition of an error-detecting protocol, the data transfer speed is reduced. This type of transfer would result in an 8-bit byte of information being sent at a rate of 8 multiplied by 1/9,600 seconds or 832 μs. For ISDN applications, data can be transferred at 64,000 baud or one 8-bit byte every 125 μs. If a microprocessor is used to transfer these data on a byte-by-byte basis, then the amount of overhead used by the transfer would be $(t \setminus 125) \cdot 100\%$, where t is the time required to transfer one byte in microseconds. Two channels would more than double this overhead.

There are other options available. For example, if a FIFO is used to temporarily store the data while it awaits transmission then the overhead for the transfer of one byte would be $([t \setminus 32] \setminus 125) \times 100\%$, where t is the time needed to transfer 32 bytes of data.

Finally for DMA transfers the overhead required to transfer 64 kbytes would be $([t \setminus 64{,}000] \setminus 125) \times 100\%$, where t is the time required to set up the DMA controller.

This overhead would be desirable for data transfers in which larger packet sizes are common. The price paid for larger packet sizes is a slower user data rate on poorer quality lines. When an error is detected in a received packet, the receiving station will request a retransmission of the packet. With a larger packet there is also a greater probability that an error will occur.

For example, suppose that the error rate on an ISDN line were 10^{-7} and the packet size were 100,000 bits long. This would result in an error occurring once in every 100 packets. So, to transmit 100 packets successfully, 101 packets would have to be sent; therefore, the effective data rate would be 100/101 times the actual data rate. If a 10,000-bit packet size is chosen, then a user data rate of 1,000/1,001 times the actual data rate would result. Even though DMA architectures can result in the unattended transfer of large packets, this may be undesirable if the error rate on the line is high.

The data channel in ISDN is controlled by signaling information across the D channel. This means that there should be two distinct paths through a terminal for the data channel: one for the actual data itself, and one for the flow control of the channel. The control of the channel could involve interaction with another process in the terminal or interaction over an external interface (RS232, for example). In many instances, device manufacturers will provide serial interfaces

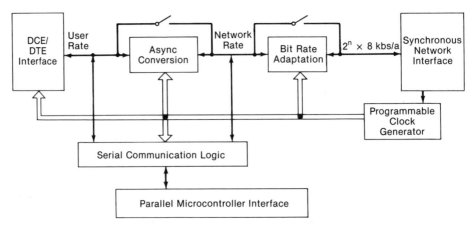

Figure 4.11a. Functional Block Diagram of Siemens PSB2110 ISDN Terminal Adapter Circuit (ITAC).

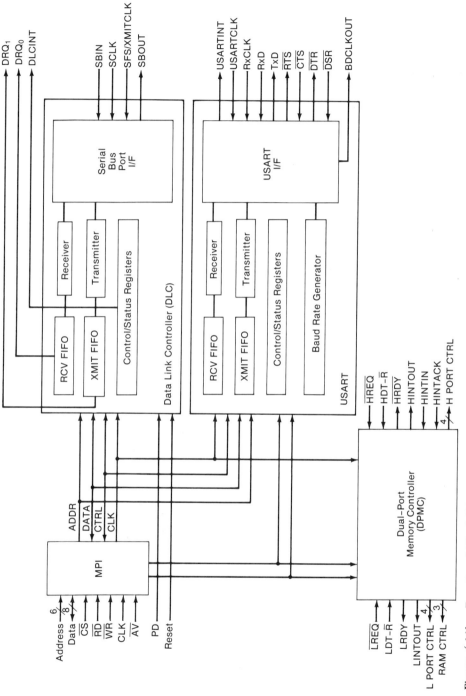

Figure 4.11b. Functional Block Diagram of AMD Am79C401 Integrated Data Protocol Controller (IDPC).

between components to transfer the user data and a microprocessor interface to transfer the flow control information.

In certain applications the data formatting can be performed entirely by an application specific device. Two types of solution can be used to provide this type of function. One is a specialized device to interface a terminal to an ISDN device; the second is a microcomputer with a program to format and control the data. Depending on the availability of specialized processors for the task of data transfer, either of the solutions may be chosen. Most of the interchip strategies used by device manufacturers facilitate interfacing a microcomputer chip; such as the 8051 family.

Two approaches are shown in Figures 4.11a and 4.11b: one using a DMA approach to control the data transfer, and one using a dedicated rate adaption scheme (CCITT V.110).

As with call control procedures, the choice of which protocol or rate adaption method to use depends upon the application. There are several types of protocols currently in use and proposed for data transfer over ISDN. Additionally, there are proposals to support packet switching in the network. In some applications it is desirable to have the hardware and software available to cover several different types of data transfer schemes.

The Voice Interface

An integral part of most ISDN terminal designs is the voice interface. This interface provides two functions: digitization of analog voice signals, and control of telephone functions. The digitization of analog voice signals is not new. The main difference with an ISDN network is that the digitization is performed at the terminal instead of at the line card in a digital switch architecture. This new level of functionality increases the complexity of the design task. The availability of a digital voice channel, however, does have its advantages. Voice mail services can easily be added to an ISDN telephone. Alternatively, *call progress information* (dial tone, busy signal, etc.) can be realized as voice messages from a digital memory.

The first problem that has to be overcome is that of digitizing the voice signal. In an analog line card the digitization process is different than that needed at the terminal. The line card not only digitizes voice signals but also compensates for the frequency distortion added by the telephone line across the voice spectrum. The filters that are used for this are not necessary to compensate handset performance in an ISDN telephone. Instead there a different type of filtering is needed.

Three filters are needed in an ISDN telephone: one to compensate for the performance of the earpiece; one to compensate for the mouthpiece; and one to give *side tone gain*. Side tone gain is the inherent feedback from the mouthpiece that occurs in analog systems. The performance of a telephone will seem unnatural if the people talking into the handsets cannot hear themselves in their earpieces. The voltage levels that appear at the telephone handset will be different than those seen on a line card. For these reasons, new codecs and filters are being designed to meet these requirements. Although codecs used in line cards can be used in a telephone, they are not optimized and may prove to be expensive due to the additional circuitry necessary to interface to the handset.

Another requirement of ISDN telephones is *tone generation*. In an analog telephone the tones are generated by the exchange system. In an ISDN terminal, tones may be needed for a busy signal (for example, when a third telephone tries to get a line on a point-to-multipoint S interface), or DTMF tones may be needed for stimulation of analog equipment in the network. These tones can either be generated in the analog domain or generated and added to the voice signal digitally. If the tones are generated digitally, then some type of digital signal processing is needed. The DSP can either be added as a separate device or as part of a codec.

An ISDN telephone must also generate ringing at the terminal. The ringing signal is now realized as a sequence of D channel messages and not as an AC signal superimposed on the telephone signal. In ISDN telephones this can be implemented by the use of a piezoelectric device. This allows the telephone to operate from 5 V supply rails. A ringing generator will have to be included as part of the telephone design to drive the ringing device.

An example of a circuit that has been optimized for ISDN telephone applications is shown in Figure 4.12.

An alternative solution to providing voice service in an ISDN terminal by using a special telephone is to add a telephone adapter to the design so that a standard analog telephone can be powered. In effect this involves moving the line card interface of the analog telephone found on exchanges into the ISDN terminal. A few unexpected problems may be encountered.

The first of these is layout design. The ringing signal needed to operate a standard telephone is a comparatively high-voltage AC sine wave (in the order of 60 V or more). This type of signal can cause many noise problems in the design if not handled properly in the layout. Fortunately, many *analog line interface circuits* contain a synchronization input that synchronizes the relay controling the ringing signal to the zero crossings of the sinusoidal ringing signal. This ensures that the ringing signal is switched on and off the line when no current is flowing into the telephone ringer and thus minimizes the induced switching noise.

Both the ringing circuit and the telephone itself must be powered. If the telephone is to be powered from an ISDN terminal, the appropriate power levels must be made available. This could prove difficult when interfacing to a PC. To do this a DC-to-DC converter is added to the terminal design to change the voltage levels from the PC supplies to the levels for the telephone application. Normally these are switching power supplies. Again, care must be taken both in design and layout to ensure that minimal noise is injected into the voice channel and digital line interface.

In many instances this type of adapter is used as part of a voice/data terminal in which it may be desirable to allow the telephone to operate even though the data portion of the terminal is not in use and not powered. The equipment is then powered from the ISDN interface. This is the subject of the following section. An example of this type of adapter is shown in Figure 4.13.

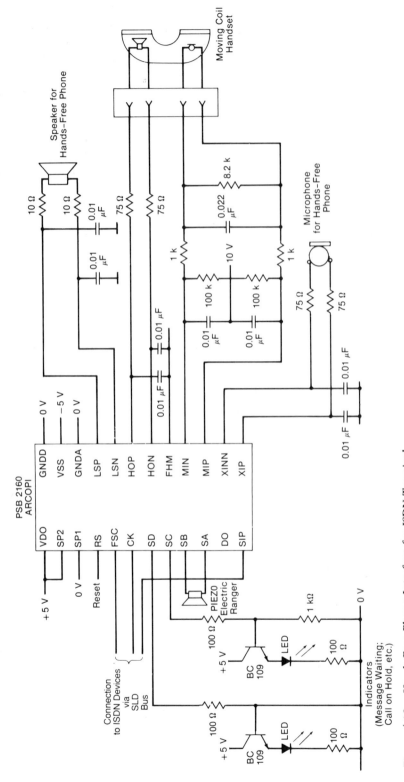

Figure 4.12. Hands-Free Phone Interface for ISDN Terminal.

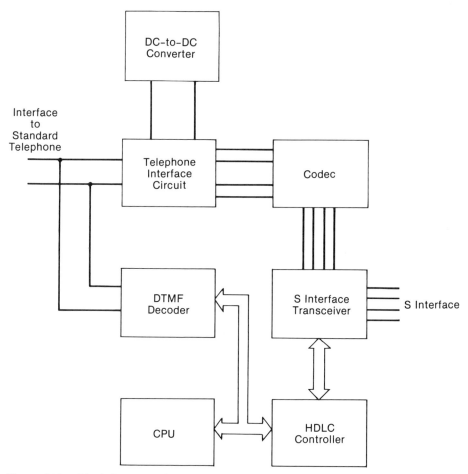

Figure 4.13.   Block Diagram of a Telephone Adapter.

Power Supplies for ISDN
Terminal Equipment

There are three different methods of supplying power to ISDN terminals as defined by CCITT I.430. These are *local*, *direct*, and *phantom* power feeding. The different types have distinct limitations and each requires unique circuit designs.

Perhaps the easiest of the three methods to understand is local powering. This implies that all power is derived from a local power supply, for example, a 110 V line supply. Although the design of the supply can be easily achieved, there is one unique twist in the design of an ISDN terminal. When the TE is not powered and still connected to the line, it will present a load on the transmission line. This is particularly important in point-to-multipoint applications. If the S interface cannot meet the CCITT impedance templates when no power is available to the terminal, then a relay must be added between the line and the S interface of the terminal. This relay will disconnect the terminal from the line when no power is available.

The second of the three modes is also easy to understand. Direct power feeding, as the name suggests, is a scheme whereby power is supplied from the exchange over two separate lines. One part of this type of power feeding circuit would be a regulator in the terminal, as the voltage level seen at the TE will depend upon the ohmic loss across the telephone line to the exchange. As the length of this line varies, the level seen at the terminal will also fluctuate.

The third type of power supply is phantom feeding. This approach uses the lines carrying the ISDN traffic to supply power. This cuts down on the number of lines needed to connect the telephone to the exchange. Two types of power feeding arrangements are shown in Figure 4.14: one for a four-wire S interface and one for the two-wire U interface. This type of feeding method has the additional advantage of providing a *sealing current* for the telephone line. A sealing current is a small DC current that reduces the oxidization of the telephone connections on the line.

Because the line voltage provided in phantom feeding is quite high—between 20 V and 48 V or higher—a DC conversion must be performed together with regulation of the output voltage. This is normally provided by a *regulated switch*

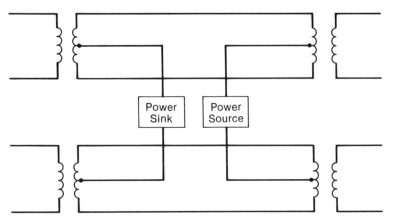

Figure 4.14a.   S Interface Power Feeding Arrangement.

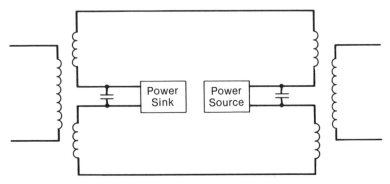

Figure 4.14b.   U Interface Power Feeding Arrangement.

mode power supply circuit. These circuits rely on the *back EMF* generated by a transformer in an oscillator circuit. There are many such devices on the market to design this type of function. One important criterion must be met. The phantom mode can only supply relatively small amounts of power, so the power supply circuits must consume little power themselves and have a high conversion ratio (power out divided by power in). This virtually makes the use of CMOS technology mandatory.

A switched mode power supply, as outlined, uses an oscillator circuit to step up the DC voltage. This oscillator circuit can insert a considerable amount of noise into the line interface. Two measures can be taken to combat this interference. One is to add inductors, *high-frequency chokes*, between the switching power supply circuit and the line interface transformers. The other is to synchronize the frequency of oscillation to the clock derived from the incoming data stream. This latter measure will insure that any noise picked up at the line receiver will occur in the input pulse transition and not affect the sampling of the data stream. In addition to these specific actions, general rules of layout, grounding systems, etc., should be observed.

When phantom power feeding is used, the line interface transformer plays an important role in the system. Because the power is supplied differentially along the line, any variances between the resistance of each of the paths will result in a DC offset on the data signals. The DC offset can alter the pulse shapes and reduce the opening of the eye diagram and increase the sampling errors. The ability of the transformer to present equal paths to the power supply on the line is called the longitudinal balance. Limits for the longitudinal balance are given in the respective line specifications.

When the exchange loses power and has to resort to local battery power, then the total amount of available power for the subscriber equipment has to be reduced. This is signaled to the TEs by reversing the supply voltage polarity. Only equipment that can operate at the reduced power levels can be left connected to the power supply. This type of equipment is known as a *priority terminal.* The priority terminal must be able to detect the voltage reversal and reduce the amount power taken from the line. In the case of the S interface this is limited to less than 400 mW. The other terminals can be left on standby but cannot take more than 25 mW line power. In a terminal that will operate in priority mode, it is quite a demanding design task to implement a circuit with microprocessor, memory, codec, and line interface that will operate on a power budget of under 400 mW. This would again lead to chosing CMOS components for the design.

Future Directions of Terminal Designs

As ISDN becomes increasingly widespread, the terminals that are designed will become more complex. There are several newer developments that are changing terminal design.

Greater integration of devices helps to reduce the cost of terminal manufacture. For example, the Am79C30 digital subscriber controller (DSC) from Advanced Micro Devices has three of the functional blocks of a terminal. The DSC contains the line interface, an HDLC controller for layer 2 processing, and a codec/filter for telephone control. This higher-scale integration reduces the number of devices, reduces the cost, and increases the reliability of terminal designs.

Another example of higher-scale integration is the development of LAPD controllers. As the LAPD protocol becomes more standardized, it will be easier to implement the functional block in silicon. This removes the burden of processing the layer 2 control data and allows the microprocessor to perform additional functions.

Although there are standards set for the line interfaces, the latest versions of some of the line interface devices can exceed these limits. For example, CCITT I.430 stipulates that the range of the S interface should be 1 km. The newer transceivers can exceed this limit and operate in the range of 1.6–2 km and still maintain the error ratio required on the line.

5

The ISDN Exchange

Functional Blocks of an Exchange

The functional blocks of an ISDN exchange are to a great extent the same as in any digital switch. Because ISDN brings with it two changes—digital transmission and new services—the exchange system design must meet these enhancements. All blocks of the exchange will be affected by the need to support ISDN. However, some areas will be affected much more dramatically than others. Also, the exchange system architecture may have to be altered to optimize its operation. Before discussing the effect of ISDN on exchanges, an outline of the operation of a digital exchange will be analyzed (see Figure 5.1).

The backbone of a digital exchange is the PCM highway. This highway can have one of two forms. It can either be a serial or parallel bus structure. The advantage of parallel structures is that the bus transfer rate is much higher. The advantage of serial structure is that the number of interconnect signals is reduced. With newer interface devices, PCM serial highways can achieve data rates in excess of 8 Mbs. Care must be taken in designing the backplane hardware to realize such a bussing structure. When using this type of high-speed serial bus, attention must also be paid to the interaction between clocking and data signals. For example, it may be necessary to delay the recognition of a start of a PCM frame from the framing signal. This can be required to the different path lengths of the backplane interconnect. PCM timing signals are normally routed to all cards on the highway. If the path for data and timing signals is different, then the timing relation will change and a hence a delay must be added. The use of slower-speed parallel busses does not remove the effect but does, however, minimize it.

Each functional block within the exchange uses the PCM highway to transfer

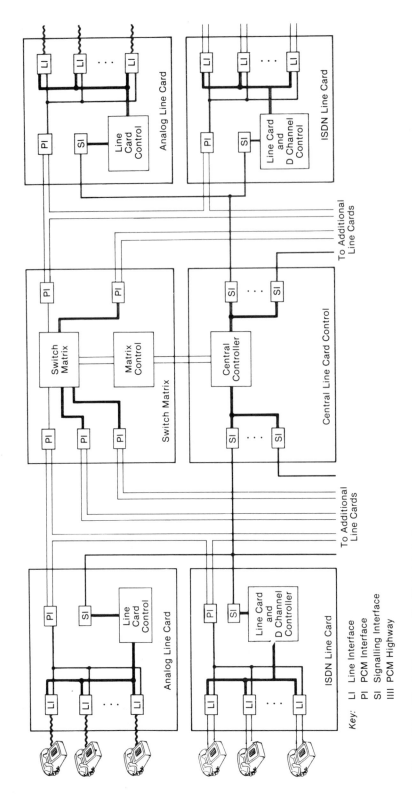

Key: LI Line Interface
     PI PCM Interface
     SI Signalling Interface
     |||| PCM Highway

Figure 5.1. Block Diagram of a Digital Switch.

information. This information can either be voice/data information from the telephone line or signaling data passed between blocks. In many cases separate data paths are used for the two types of traffic. In this way each of the major parts can route telephone information through the exchange from one line to the next. Because a PCM highway will support a number of channels (dependent on the transfer speed), then the number of highways required will depend upon the number of subscribers the exchange will support. The majority of exchanges will have at least two PCM highways, and most probably more for larger exchanges so that a reasonable number of subscribers can be supported.

In any digital exchange there are four major functional blocks: line cards; line card controllers; switch matrix; and power supply.

The line cards, line card controllers, and the switch matrix will all use signaling PCM highways to communicate. Only line cards and the switch matrix will use PCM highways that transfer telephone information.

The Line Card

As the name suggests, the job of the line card is to provide an interface between the telephone line and the PCM backplane. This involves four major tasks: digitization of analog telephone signals; interfacing to a time division multiplexed PCM highway; line card control; and communication with the line card controller. It should be pointed out that with advances in codec/filter design, it is possible to have the circuitry to interface to several telephone lines on one printed circuit board. Moreover, with the use of *surface-mount devices*, it is possible to have interface circuitry for as many as 32 telephone lines on one line card.

The main purpose of the telephone line interface is to convert analog telephone information into a digital form and vice versa. The line interface also provides power via the line to the telephone equipment. In line cards that interface to telephone lines that go off-premise, additional protection circuitry is added to ensure that any extraneous voltage appearing on the line does not damage the line card circuitry. The line card will also contain the necessary circuitry to detect the condition of the hook switch in a subscriber's telephone.

The PCM highway interface will concentrate the serial highways from the codecs and switch them into *time slots*. This involves some sort of programming because time slots are normally allocated dynamically. In other words, a time slot for telephone information will be chosen when the call is established. Once the call has been completed the time slot is returned to the pool of free time slots. To accomplish this, control information is transferred between the line card controller and the interface to the PCM highway. This can be either with or without the aid of a microprocessor on the line card. The PEB2050 peripheral board controller from Siemens and Intel can be used to configure the time slot allocation remotely. In addition to time slot allocation, the interface should provide some clock shifting for the PCM highway. This will solve the problems caused by incorrect alignment of the framing pulse and the PCM data on the exchange backplane.

The peripheral board controller is one solution to the signaling need between the line card and controller. In more complex line card designs (for example, ones that can support a feature such as *third-party conferencing*), a different signaling scheme is used. These enhanced schemes require the use of a microcontroller on the line card. In quite a few cases the signaling system that is used is HDLC, so an HDLC controller with a PCM interface handles the link to the controller from the line card.

The local processing power on the line card can also be used for other functions. The standards governing telephones throughout the world vary considerably. To meet these variations, codecs with programmable DSP filters are used to match the line conditions. A local microcomputer can be used to control such codecs.

The Line Card Controller

The line card controller interfaces between the switch matrix and line cards. This functional block will normally be responsible for controlling a group of line cards. The controller will not normally have access to the telephone channels in the switch. However, the line card controller may use the channels for control purposes. In a digital switch it is possible to have only one call progress tone generator for all of the lines. This generator is allocated a permanent time slot within the system. When access to a call progress is required (for dial tone, etc.), the switch matrix connects the telephone line time slot to the call progress tone generator time slot. The dial tone digital pattern is simply "copied" to all of the connected channels. In practical terms this will mean that the same speech memory address will be programmed into several connection memory time slots.

Another function that is normally performed centrally is DTMF recognition. A telephone channel is switched to a central DTMF processor that decodes the dialing information and then passes it to the control circuitry of the switch matrix. The switch matrix will then disconnect the channel from the DTMF processor and route it to another line card, or even the call progress tone generator if the line is busy.

The Switch Matrix

The function of the of the switch matrix is to transfer telephone information from one line card to another. In a digital exchange this is done in two ways. The first way is to switch the information from one PCM highway to another. This is

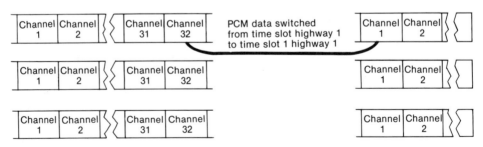

Figure 5.2a.   Example of Time Switching.

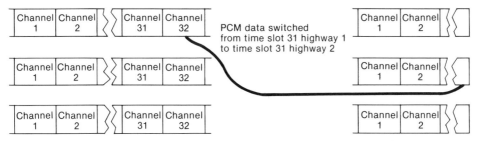

Figure 5.2b.   Example of Space Switching.

known as *space switching*. The second is to move data from one time slot to another, which is known as *time switching*. Often both time and space switching are performed simultaneously.

There are a few devices available that are dedicated to the task of time/space switching. The operation theory of these devices is based on the concept of dual port memory. The serial data are first converted to a parallel format; the information from the PCM highway is then written into sequential locations of the *speech memory*. The information is then read out of the memory, serialized, and output on the PCM highway. The sequence in which the information is output is determined by the contents of the *connection memory*. The connection memory contains the output information addresses in the speech memory.

For example, the information on PCM highway 1 in time slot 0 would be written into speech memory location 0. However, the contents of the connection memory in location 0 would give the address of the information in the speech memory to be output on PCM highway 1 time slot 0. The input in speech memory location 1 might not be output until, say, PCM highway 3 time slot 10 (see Figures 5.2a, 5.2b, and 5.2c). In most cases though the link through the switch matrix is bidirectional.

As with the line card, there must be a signaling link between the line card controllers and the switch matrix. The line card controller can request a path from the switch matrix. Once a path has been programmed into the connection memory, the line card controller can signal the configuration information to the line card. Again HDLC signaling can be used for this link. Because many paths are created and dissolved by the switch matrix in real time, a powerful microprocessor is normally used to control the programming.

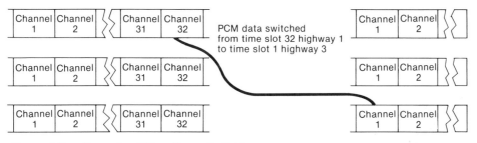

Figure 5.2c.   Example of Time/Space Switching.

Effects of ISDN on the Digital Exchange

ISDN will alter both the individual functional blocks and the architecture of digital exchanges. Each of these changes will be dealt with in relation to the functional block descriptions outlined previously.

The ISDN Line Card

The functional block that will be most affected by ISDN is the line card. The line interface in an ISDN will be of a digital rather than analog format. Although this will solve some of the problems that face analog line card designers, it will also add new challenges.

In an analog line card, there is no absolute necessity to include any form of local microprocessor control. The control function can be passed back to the line card controller. As with an ISDN telephone, even the most rudimentary line card design will need a microprocessor. Processing control is needed to control the line interface functions and in some cases the higher levels of the protocol interaction. There are several different types of system architecture that can be used to realize the control of line card functionality.

Depending on the degree of protocol processing and the number of lines supported, the type of processing power required on a line card can vary. If only physical interface control is performed, then a relatively simple microcontroller can be used. If, on the other hand, full processing of all layers of the protocol software is performed for several lines simultaneously, then a sophisticated processor system is needed. An HDLC controller would be virtually mandatory for processing the lower levels of the protocol for call establishment. The addition of these components causes two difficulties: first, extra space is needed for the supplementary devices; second, extra software is needed for protocol control.

The addition of extra components is not only an additional problem for board layout, but also for power dissipation. In many exchanges with numerous line cards, power dissipations can cause serious headaches. In the case of ISDN line cards, it is vital that a low-power technology, such as CMOS, be chosen for the interface and HDLC circuitry. For applications in which protocol processing is done on the line card, a powerful processor must be used. Unfortunately "powerful" sometimes means not only a high degree of processing capability but a high power dissipation as well.

Anything that reduces the need for a highly sophisticated microprocessor will solve several problems at once. The inclusion of FIFO memory on HDLC controllers can play an indispensable role in reducing the amount of microprocessor intervention required for protocol processing. In call control protocols, the message size is normally quite short (less than 32 bytes), so a small FIFO size can easily handle complete messages. This can get around difficulties encountered in allocating memory for incoming and outgoing D channel messages for ISDN telephone lines.

Another solution to the problem of processing call control protocol is the use of DMA. With the small message size and the need to incorporate a complex DMA controller to handle all the line interfaces, this solution may prove to be unwieldy and costly. Another point to be looked at in a DMA solution is the power budget. DMA controllers usually dissipate a high amount of power.

One problem that is carried over from analog line card designs is that of layout. Although the telephone interface is digital, it is still susceptible to noise on the line interface. Therefore, ground planing should be carefully reviewed. Fortunately, because analog line cards have been in production for a few years, there is a great deal of experience in minimizing the telephone interface.

All of the hardware issues that affect ISDN telephone line interfaces that were discussed in chapter 4 apply equally to the line card, with exception of those issues related to the multipoint operation of the S interface. There will always be only one network termination on the S interface. The transmitter must be able to drive the loading of eight receivers in the case of a point-to-multipoint S interface.

For the U interface, the issue of protection circuitry is more critical than for the S interface because the U interface is intended for off-premise service. A full protection circuit is essential for U interface line cards. The protection circuit should not interfere with the operation of the U transceiver. Because the interface for a U reference point is meant to be point-to-point only, there is no need to be concerned about operation of the protection circuitry without power. A normal circuit with diode clamps to the power rails plus a discharge device will suffice for the purpose of protecting the U interface.

The last part of the line card to be affected by ISDN is the PCM highway interface. ISDN supports both voice and data; in addition, the call control protocol can establish which type of call is requested at the start of a call set-up procedure. Because of this, some exchange architectures use two PCM highway systems: one for voice and one for data. The line card must have the ability to support the two highways in such a system. The two telephone PCM highways and the signaling highway mean that the line card must support three PCM highways in total.

Given the case in which only part of the call control protocol is supported on the line card, the rest of the processing must be performed by the line card controller. This implies that the call control messages must be interspersed with the line card control messages. If the line card control signaling system is not HDLC based, this can cause considerable processing problems on the line card when multiplexing the two protocol systems together.

The ISDN Line Card Controller

As shown with the line card, the effect on line card controller design is dependent on the partitioning of the D channel call control protocol processing. The degree of difficulty in implementing this partition is dependent on the allocation of processing between the line card and the controller and also on the existing method of communication between these two functional blocks. The impact of existing signaling systems cannot be ignored, as many digital exchanges are constructed on a modular basis. This means that there could be a mixture of ISDN and analog line cards within the same exchange. The signaling structure for the analog line cards must therefore coexist with any new system employed for the ISDN line cards.

Additional functionality may be needed in the line card controller. For instance, in systems that differentiate voice and data calls, a method of detecting and acknowledging different calling procedures must be implemented. This is particularly true if the exchange has a mixture of ISDN and analog line cards.

ISDN could have the impact of removing call progress tone and ringing generators. In a fully ISDN exchange, there will be no need for a ringing generator because the ringing is signaled to the terminal via a D channel message. It is arguable whether the need for call progress tones will disappear in an ISDN exchange. This will mainly be a function of whether the terminal equipment can generate these tones.

The ISDN Switch Matrix

Probably the least affected of all of the exchange functional blocks is the switch matrix. The two main effects that ISDN will have on the matrix will be a necessity to double the switch matrix so that one can handle voice and one data; and, particularly at the primary rate speeds, there is a need to concatenate B channels.

When voice switching is augmented by a second switch matrix for data, the main concern will be in the logistics of adding extra hardware. There should be little impact on the actual design of the switch matrix itself. Even the software could be similar. The major advantage of this type of architecture is the ability to easily route the data to an external processor for packet switching. These additional services can be added because of the integration of data into the telephone network.

Although the data rate on an ISDN line is far in excess of that of a conventional analog line, for some applications even the 64 kbs^{-1} is not fast enough. Video services are such an application. For these types of higher-speed applications, there is a need for a 128-kbs channel to be switched through the exchange. This would mean that the software in the switch matrix would have to be upgraded such that two 64 kbs time slots on the PCM highway can be switched through as if it were one channel.

Packet Handling in an ISDN Exchange

Due to the digital nature of ISDN, some architectural changes have to made in exchanges. The functional blocks will have to deal with packetized information over the ISDN line (see Figure 5.3a–c). In current digital exchanges, the various blocks communicate with each other using a signaling system over the PCM highways.

There is no reason for a digital exchange manufacturer to use HDLC protocol for signaling between the functional blocks of the exchange. In fact, the use of HDLC may be considered too complicated a solution for this task. The system

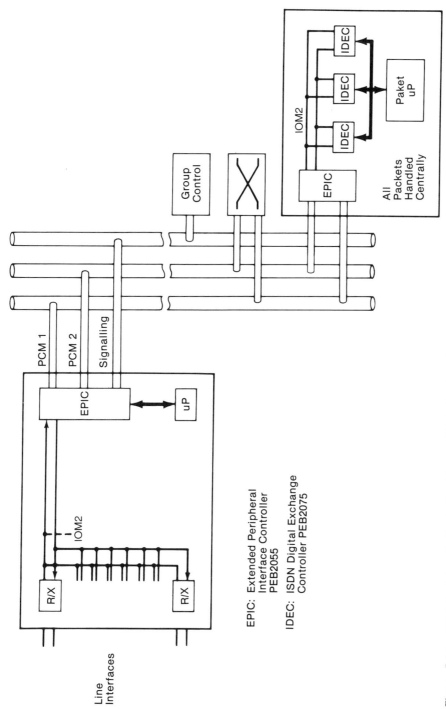

Figure 5.3a. ISDN Switch with Centralized Layer 2/3 Packet Handler.

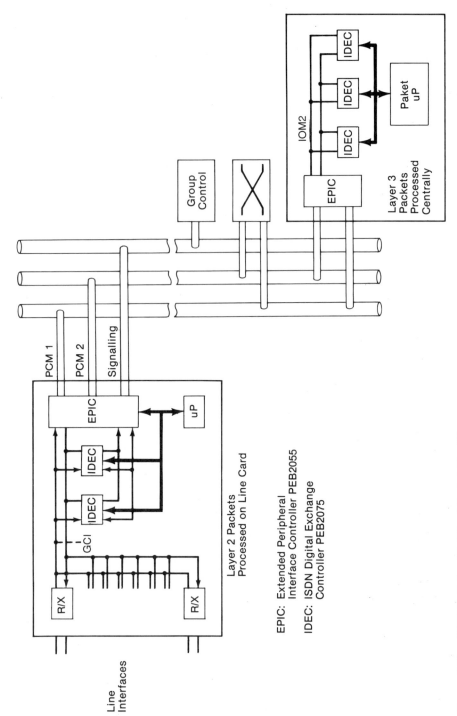

Figure 5.3b. ISDN Switch with Mixed Packet Handling.

92

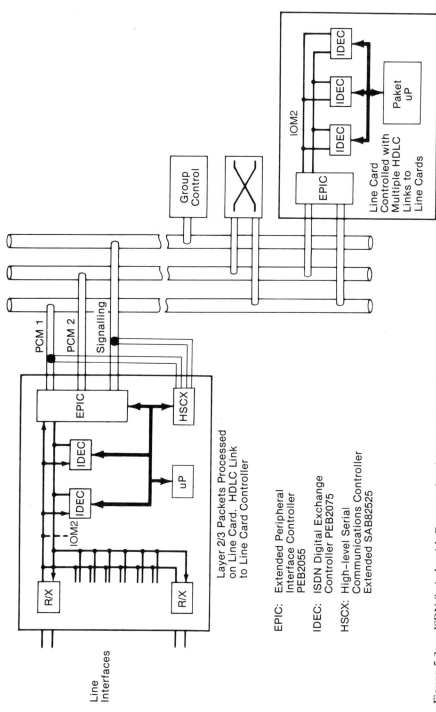

Group Control

PCM 1
PCM 2
Signalling

EPIC

IDEC

IDEC

uP

HSCX

IOM2

R/X

R/X

Line Interfaces

Layer 2/3 Packets Processed on Line Card. HDLC Link to Line Card Controller

IOM2

EPIC

IDEC

IDEC

IDEC

Paket uP

Line Card Controlled with Multiple HDLC Links to Line Cards

EPIC: Extended Peripheral Interface Controller PEB2055

IDEC: ISDN Digital Exchange Controller PEB2075

HSCX: High-level Serial Communications Controller Extended SAB82525

Figure 5.3c.   ISDN Switch with Decentralized Packet Processing.

chosen may not conform to the lower three layers of the OSI model because many digital exchanges were built before the general acceptance of this model. These internal signaling schemes can cause considerable difficulties for the ISDN line card designer. The main problem with regard to the signaling issue is how much processing is done on the line card.

The first solution is to perform all protocol processing for the lower three layers on the line card. The input and output primitives to the layer 3 of this protocol are converted into whatever signaling system is used in the exchange. This at first sounds like a simple task; however, the amount of processing required for an ISDN line is considerable. To implement simple POTS across ISDN, about 64 kbytes of code need to be executed per line. In line card architectures that handle many lines with one card, an extremely powerful processor would be required to perform this task.

The power of the processor could cause problems on the line card. One problem is the increased space needed for the processor and the support hardware. Another is the high-frequency clock that is needed by such devices. This could interfere with the digital line inputs. Yet a third concern is the fact that the processor and peripherals would use an appreciable amount of power from the backplane power supply, leading to distribution and dissipation difficulties.

An alternative solution to the approach of ISDN protocol processing on the line card is to use a multiprocessor design. This would involve using a less powerful processor to handle layers 1 and 2 of each ISDN line, and then one single processor for layer 3 and internal signaling. By using this type of design, the processing power is distributed throughout the line card, allowing for less sophisticated devices to be used. By using simpler processors or even microcomputers, less support hardware is needed; these types of devices are available in CMOS.

Another architectural approach is to only process layers 1 and 2 of the ISDN protocol on the line card. Layer 3 is then processed centrally. This has the major advantage that the amount of processor capability required on the line card is substantially reduced. The difficulty of this technique is integrating the layer 2 and 3 messages from the line card with the existing internal exchange signaling system. Mixing an HDLC-based protocol with a non-HDLC one can preclude the use of dedicated HDLC controllers. For example, the flag pattern for HDLC, 0111 1110, could occur in the signaling system of the exchange. When this pattern occurs the HDLC interprets this pattern as a flag. Some method of appropriating the PCM highway to differentiate its use must be established so that the transmission of ISDN HDLC packets and internal signaling data can be interleaved.

The last approach is to remove all protocol processing from the line card and place the burden onto the line card controller. This fits into the architecture of existing digital switches in which only line control functions are performed on the line card. This has the same drawback as the previous architecture. In addition, the amount of traffic on the PCM highway is increased as all layer 2 packets are now transferred on the highway. Because these packets have to go through extra transformations from the line to the PCM highway and then to the layer 2 and 3 controller, the roundtrip delay will be increased.

The advantage of this type of architecture is the substantial reduction of the amount of processing power needed on the line card. A microprocessor will still be required to control the layer 1 functions on the ISDN line. However, this processor could be a simple microcomputer.

The change in the types of ISDN devices helps reduce the problems of ISDN line card design. There are two routes that suppliers of ISDN devices are taking. One is to analyze these architectures and partition the functions on a specifically designed bus structure to minimize the problems of processing and device count. The second is to integrate more dedicated higher-level functionality into controllers to handle more protocol processing.

The GCI bus mentioned in chapter 4 has been designed to help alleviate the difficulties of producing line cards for ISDN applications. Because the bus structure is modular, it is suitable for different line card designs. The ability of the bus to allow the B and D channel information to be passed along the same electrical interface makes it applicable to all of the three line card architectures. There are devices being produced now that support the GCI bus structure for line cards. Siemens, being one of the pioneers of the bus, has several devices. These range from the line interface for either the U or S reference points to the PCM interface and a series of HDLC controllers.

Other bus structures are similarly supported by their manufacturers. Motorola has a series of devices to support line cards based on the IDL bus. These include line interfaces and a dual HDLC controller. The B channels can be passed through to the PCM interface by time slot assignment.

AT&T also provides devices to support its bus structure. In addition to supplying the line interface and an HDLC controller, AT&T also has a multichannel HDLC controller that is suitable for high-density line cards. This is one example of higher density integration minimizing the design constraints facing the ISDN line card designer. The SPYDER-S can handle up to eight HDLC channels. This part integrates other functions on-chip to facilitate line card designs. Integrated onto the device are eight transceivers, a 16-channel DMA controller, and an interrupt controller. This device can be used to control as many as eight line interface transceivers—for instance, the ISDN user interface for switches (UNITS) T7252.

Another solution being put forward is higher functionality in the HDLC controller. Most HDLC controllers include the 0 bit insertion, flag control, CRC generation/checking, and address recognition. Some even include basic functions for layer 2 control. Some manufacturers are taking the functionality of the controller one level further by including full layer 2 processing capabilities.

The HPC16400/HPC36400/HPC46400 family of high-performance microcontrollers is customized by adding two HDLC controllers to fit into ISDN designs. By integrating these two parts of a line card, the device count can be reduced. The task of writing a program to control the microcontroller still remains, and this could pose some additional problems. From a hardware perspective, the main impact this would have is that multiple copies of the program would have to be contained on the line card. This would give rise to large amounts of read-only memory (ROM) on the card. For a simple implementation of layer 1 and 2 functions at the line card end, about 20 kbytes of code are necessary. Because each microcontroller has to have access to its program at all times, eight copies of the program must be placed on the card; a total of 160 kbytes of ROM would be needed. So instead of having one ROM for code space there is now an eightfold increase. In other words, the price that is paid for reduction in the processor functions is an increase in devices for ROM code.

Semiconductor manufacturers are becoming increasingly adept at integrating larger numbers of transistors on a single chip. This trend heralds the development of a single device that will perform all LAPD functions. Whether this will be desirable from a cost versus complexity standpoint is debatable because line cards are a very cost-sensitive area of the exchange.

A move toward higher-density HDLC controllers is being pursued to an even greater degree by some manufacturers. Controllers with 32 channels are now available to address the line card applications. The current implementations do require the user to follow specific system designs to make full use of these devices. As with the addition of the microcontroller to the HDLC controller, high-density HDLC controllers may require an unacceptable amount of "glue" logic to fit them into a design. Again this may make them unusable from the cost-sensitivity view point.

Data Processing in the Exchange

Data packets will also be a part of the ISDN environment. How the exchange deals with this technologically is another area of ISDN's impact on the system. Forgetting for a moment the legal side of the problems associated with data handling, it is worth looking at some of the possibilities that ISDN offers the exchange manufacturer. There are two types of application that the exchange manufacturer can address. One is packet switching of data and the other is protocol conversion.

Because the ISDN exchange now has to address the handling of packetized information to control the subscriber equipment, the development of the software and hardware for packet data is a logical progression. The packetized data can either come over the B channel or be packet switched on the D channels. If packet switching is to be performed in both cases, then the capability must be added to the exchange.

Extra capability in the system architecture to route the packet data can be given to line card controllers. This would involve processing layer 3 data call networking information. This could call upon applications ranging from selecting which outgoing telephone line the data packet is destined for, to performing a *statistical multiplexing function* over a shared link. Packet switching can be used for rate adaption between non-ISDN equipment and ISDN equipment. Therefore, the exchange must be able to handle the rate adaption protocols if any interaction with the data is to be considered. This would be particularly true of statistical multiplexing.

As ISDN is introduced, there will be an interim period during which it will be necessary to include an analog transmission path within the network. One example of this is *modem pooling*. Because it has taken more time to develop the U interface standards than those for the S interface, there will be a boom in applications using the S interface. To solve the problem of transmitting data over a WAN, a technique of modem pooling can be used. In this technique, several analog modems are connected to the off-premise side of a PABX. The exchange can then allocate the use of the modems for data transfer as demand dictates. Some

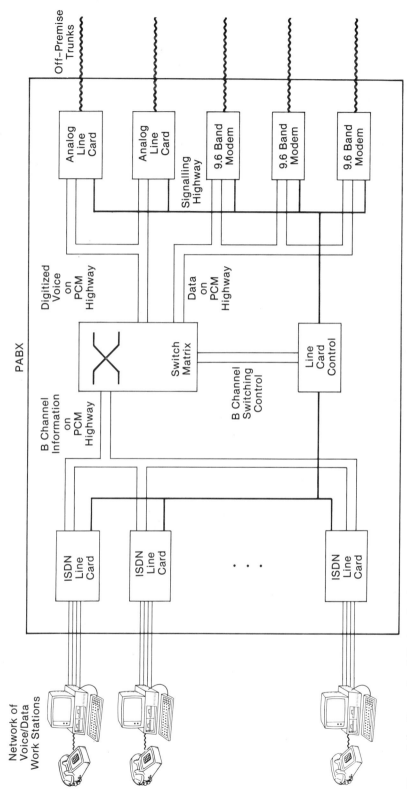

Figure 5.4. Example of Modem Pooling.

schemes have been proposed such that several lower-speed modems can be linked together to provide a data path for ISDN 64 kbs data. The higher-speed packets are split up and transmitted over the different modems. As one modem completes its transmission it informs the PABX that it is ready for the next packet. It is the responsibility of the PABX to provide memory space to temporarily buffer the ISDN packets while they are awaiting transmission over the analog modem link (see Figure 5.4). In fact some have suggested that ISDN will provide a short-term market for analog modems to address this type of application.

ISDN has provided the telecommunications industry with a set of transmission standards. These standards have evolved from those already in use. These earlier standards are still widely used so there is a strong inclination to incorporate them into ISDN. On the other hand, ISDN has resulted in different standards that are targeted toward the same application area. Data transfer has a myriad of different standards, both in current use and proposed applications. To enable ISDN to gain universality in the data arena, protocol conversion must be performed within the network. The exchange is the optimal position to provide this service.

Part of the call establishment procedure is a request for the type of service that the ISDN channel is intended to support. As the call is established, the called station is queried to determine if it can support the service requested. In the case of data, this may include not only the data transfer speed but also the type of protocol used. Therefore the exchange has all the information it needs to be able to select the type of conversion, if any, to allow the two stations to communicate.

A simple example of this type of conversion would be to allow a terminal using X.25 and one using the protocol sensitive mode of the V.120 standard. As these are both packet data protocols, the exchange can provide the protocol translation between the two ends. Perhaps a more widely used example would be for communication between ISDN terminals and LANs. In this case the LAN can be integrated into a WAN by providing the translation of the data protocol used on the LAN and the ISDN terminal equipment. This type of function is often referred to as a *gateway*. This type of protocol conversion has been tested in many of the ISDN field trials and found to be a viable method of integrating relatively slower-speed telecommunications equipment into higher-speed LANs.

By providing data services, the exchange, particularly the PABX, can become part of its own LAN. This LAN could link PCs, printers, and facsimile machines (faxes). This type of networking would not replace the traditional LAN but rather augment it. By giving access to the *public network*, a LAN can be made more powerful. On the other side of the coin, the data rate of ISDN is considered too slow for certain data transfer functions—downloading programs and operating systems to workstations, for instance—that are currently addressed by LAN. The goal is integrate both types of applications using *broad-band ISDN* that would employ fiber optic cable to give the extremely high data rates needed.

Power Distribution

In Chapter 4, reference was made to the power feeding configurations outlined by the CCITT standards. In an analog telephone network, the exchange provides power for the majority of subscriber equipment. For analog telephones, the amount of circuitry that is powered in this fashion is very small. This is not true for ISDN. This is because the telephone has to be a more complicated instrument for ISDN, and data terminals must be connected to the line.

Two types of exchange power feeding are outlined by CCITT recommendation I.430. One is direct power feeding and one is phantom power feeding. In both, the requirements from the exchange are the same, that is, to provide a constant voltage to the telephone line interface. This voltage should be regulated such that a constant level is supplied to the line as demand fluctuates. One further requirement is the ability to detect and protect against short-circuit overloads. These can occur either through electrical shorts due to terminal equipment being incorrectly plugged into the line, or through a virtual short circuit due to the initial current drawn as the TE powers up. In both cases the regulator must shut off the supply for a brief period and then try to reestablish the power supply to the line.

Devices are available to perform these functions. These vary from general-purpose regulators to specifically designed integrated circuits. The problem with generally available regulators is that they are not targeted to the unique environment of telephony. The additional circuitry needed to provide the short-circuit protection may be costly from both economic and space perspectives. The devices that are designed specifically for ISDN power supply applications are optimized for this task. Normally the circuitry to control more than one telephone line—typically four—is included in one chip. This reduces both cost and board space. Another factor is that most of these devices are made using CMOS fabrication.

An ISDN line is controlled by a sophisticated sequence of protocol events that must be controlled at all times by the microprocessor. It is therefore advantageous that the power regulation circuitry be able to alert the microprocessor of any changes in the line conditions. If a power glitch occurs on the interface, this will affect the line connection status. The microprocessor can then decide what action needs to be taken after the power supply has been stabilized.

If there is no requirement to supply power to the line, that is, if all terminal equipment is locally powered, then the microprocessor can control the regulator and turn off the power supply to the line. Other monitor and control functions can be included in the power regulator device. As more ISDN exchanges are built, the need for these functions will be incorporated into the devices.

A condition of emergency power distribution can occur during the operation of the exchange. An example is the case in which the local power fails for some reason. This case is covered in the I.430 recommendation, which stipulates that the exchange must go into emergency power mode. In this mode, the exchange limits the amount of available power to the S interface line to 480 mW. To signal the condition to the TE, the exchange reverses the polarity of the power supply. In most designs this would include a relay to physically switch the power supply connections to the line. This must be allowed for when designing the power supply distribution and regulation circuitry for the S interface. An example of a power supply for the S or U interface is given in Figure 5.5.

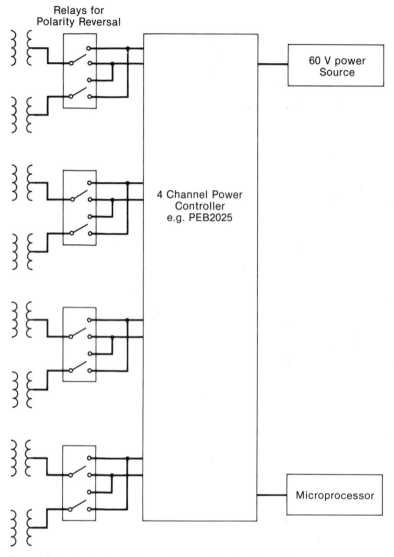

Figure 5.5.   Example of Power Feeding for the S Interface.

The S-to-U Interface

In any ISDN network there will be the need to connect between interfaces. The main area for this type of connection in the network is between the S and U interfaces. These interfaces are referred to as NTs. Depending on the level of functionality they can either be an NT1 or NT2. In the NT1, only the layer 1 functions are performed between the two interfaces, for example, level shifting, synchronization, or maintenance. In the NT2, layer 2 and 3 functions are performed in addition to the NT2 functions.

The choice of which network termination is used does not affect the level of functionality that is seen at the terminal. It does determine how much functionality is required in the exchange line card. In the case of an NT1, the exchange has full control of the S interface. This means that the allocation of the B channels is solely the responsibility of the exchange. Because an NT2 has layer 3 control, B channels can be allocated independently of the exchange. For example, this would be used to give intercom capability in a residential installation. The exchange must have the ability to query the NT to discover if a B channel is available for an incoming call.

Depending on which type of NT is desired, the amount of microprocessor power will vary. For a simple NT1 interface between the reference points little, if any, microprocessor functionality is needed. The GCI allows transceivers to be coupled "back-to-back" so that no microprocessor control is needed for an NT1 design. If an NT2 is to implemented, then a fully featured microprocessor system is needed. This would be a good application for the high integration devices in which a low-power requirement in a small space is essential. In many cases the NT will be powered by the exchange. Whether power will be provided to the S interface in this case is arbitrary.

6

Connection to the Primary Access Network

ISDN is really the culmination of the gradual digitization of the telephone network. Digital transmission of voice information has been in use since the early 1960s. The problem facing the network was its own growth: current technological approaches at the time could not support the demand. Until the 1960s each telephone link was made over a pair of twisted copper wires. As the number of subscribers grew, so did the number of twisted-pair wires. The other factor that played an important role in this increase in demand was intercontinental communication.

In the 1960s the telephone was becoming an integral part of daily life. More households began to have a telephone installed. The need to have links between cities also flourished. Having a pair of wires for every telephone link became increasingly cumbersome. The proposed solution was a system that would allow two wires to carry more than one telephone call. This would reduce the cost of installing cable between cities and countries.

Although different approaches were considered, the system that was finally adopted was the T1 transmission system. The name is derived from the identification number of the committee set up by the American National Standards Institute (ANSI). The T1 system consists of a time division multiplexed transmission scheme with 24 8-bit time slots. The time slots are multiplexed onto the same transmission medium using time division multiplexing (TDM). By using a TDM architecture, 24 telephone calls could be sent over the same two wires.

Even though advantages were gained by reducing the number of wires needed for connecting telephone equipment, the trade-off was reduced transmission distance. If 24 telephone calls are to be handled across the same physical medium, then each telephone will present an 8-bit digital representation of the voice

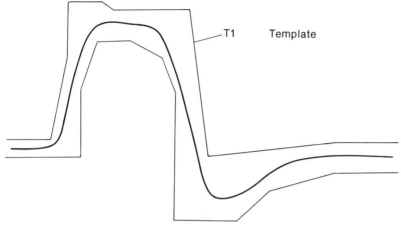

T1   Template

Figure 6.1a.   Pulse Input to Line.

information every 125 μs (this is derived from a sampling of 8 kHz to realize a 3,400 Hz voice bandwidth—see Chapter 2 for more details). This, after some simple arithmetic, results in system specification of transmission and reception of 192 bits every 125 μs, or a bit rate of 1,536 kbs. On standard twisted-pair telephone cables, the attenuation at this frequency is considerable.

To overcome this, two measures were taken. The first was to use a different type of physical medium to carry this higher-frequency transmission. Coaxial cable was installed to reduce the attenuation at these frequencies and hence increase the available transmission range between stations. The second was to include repeaters in the network. The function of the repeater is to reconstitute and amplify the digital pulses on the T1 line so that longer spans between stations can be accommodated.

Reconstitution involves restoring the pulse from the line back to its original pulse shape. A *square wave pulse* is used on T1 to represent the binary information. A square wave pulse is a combination of many frequencies, so different components of the pulse are attenuated at different rates on the line. In practical terms, this results in rounding of the pulse (see Figure 6.1). The repeater must synchronize to the incoming data and sample the line to be able to regenerate the correct binary data for retransmission.

In many earlier repeaters, the synchronization circuitry was very simple. An

Figure 6.1b.   Pulse at End of Transmission Line.

oscillator was run at a frequency close to the line pulse stream; the pulses were then used to "pull" the oscillator on frequency. This method relied upon the availability of pulses in the data stream. Because there could be many repeaters over a T1 span, the cumulative effect of a drift in frequency could cause an unacceptable amount of error in the transmission. Therefore the number of pulses, or *ones density*, on the line is important. For voice transmission this is not a problem; however for data it is a different story.

To be able to maintain the ones density, zero code suppression schemes are used. These schemes consist of replacing long sequences of zeros with a special pattern. This pattern can either be a simple replacement, as in the case of binary 8 zeros suppression (B8ZS), or a more complex method in which the zero sequence is replaced with a pseudorandom pattern, as in ZBTSI encoding. In both cases, the time to transmit the substituted sequence is equal to the time to transmit the original data, so that no degradation in the transfer rate of the information is seen.

Other schemes that are used to ensure the ones density include the inversion of HDLC data. By inverting the data the 0 insertion becomes a 1s insertion algorithm and this ensures that there are no long sequences of zeros. One limitation with this scheme, however, is that a flag idle must be used—otherwise a long sequence of zeros would result.

When considering the part played by T1 in ISDN implementations, the effect that ISDN will have upon T1 should be considered. As indicated, T1 was originally designed to transfer digitized voice information. As such certain implementations were made to aid in this application.

One of the requirements of a T1 system was to be able to transfer many telephone channels at once, and to do this 24 channels were multiplexed together. There is a special bit added to the data stream transmitted on the T1 interface. This bit signifies the start of the 24 channels so that the individual information for each telephone link can be isolated and extracted. This bit, called the *framing bit*, has a special sequence that is transmitted so that the receiver can achieve *frame synchronization*. The pattern is transmitted over a set number of 24 channel frames, and marks the start of 24 channel boundaries.

The transmission of multiple telephone calls was not the only requirement for a T1 network. The network also had to transfer signaling information. To do this, bits in the digitized voice information were occasionally used, or robbed, to transfer signaling data. This procedure of using the bits on a periodic basis is known as *robbed-bit signaling*. In a T1 system, the *least significant bit* (LSB) of each telephone link is robbed in every six 8-bit sequence that is transmitted. The bit is replaced by either an A or B signaling bit.

The bit robbing must occur on boundaries that are set on every sixth frame of the T1 line. Because there are two signaling bits, A and B, the robbing really occurs every 12 frames. The 12-frame sequence is called a *superframe* or *multiframe*. The framing bit, in addition to carrying the pattern for frame synchronization, also carries a superframe synchronization pattern. The two patterns alternate on the framing bit to allow the receiver to check both synchronization conditions.

If frame or multiframe synchronization is lost, an indication is sent to the other side. This indication is called a *yellow alarm*. If general synchronization is lost (for example, at the bit level) then a *red alarm* is sent. In response to a red alarm, the other station transmits a *blue signal*, allowing the receiver to resynchronize. The names of the alarms and signal are derived from the colors of the indicator lights used on early T1 equipment.

A 12-frame superframe is only one type of multiframing that is used on T1. A system that is becoming more widespread in T1 networks is *extended superframe* (ESF). In ESF, two enhancements are added. First, the number of frames in a superframe in increased to 24. This provides extra signaling bits for telephone signaling, a C bit and a D bit. A 24-frame pattern makes an additional bandwidth available on the framing bit. This is used to send maintenance data between T1 interfaces. More information on the T1 network is available in the many books published on the subject since T1 was developed over 20 years[1] ago.

The practice of robbed-bit signaling has been used with T1 for a number of years. With digitized voice, the fact that every sixth LSB is robbed has little influence on the overall system performance. The distortion that occurs due to the error is within acceptable limits. When data are transferred over the same network, the impact of losing the LSB every sixth byte is not acceptable. There are two solutions to get around this inherent difficulty.

Instead of transmitting data in 8-bit quantities, they are transferred in 7-bit quantities. Therefore, the fact that the LSB is robbed does not affect the overall performance because it is not used. To achieve this, data must be transmitted 7 bits at a time every 125 μs. In practical terms this involves either using a device that is capable of outputing at 7-bit increments or providing a special timing pulse to enable the data output. Many communications controller devices can be operated in the latter mode but only a few in the former. As 7 bits are transmitted every 125 μs, the data rate is reduced from the 64 kbs available to 56 kbs. In this method, 12% of the potential bandwidth is lost.

Changes in the T1 Network due to ISDN

To overcome this reduction of bandwidth, the robbed-bit signaling can be stopped altogether. If this is done, some alternative method of signaling must be used. A channel without bit robbing is called a *clear channel*. To handle signaling, the 24-channel T1 line is separated into 23 channels for subscriber use and one signaling channel. This organization of 23B+D is the *primary rate ISDN* (PRI). The signaling channel that is used between the PRI interfaces is shared between the 23 B channels and hence is called *common-channel signaling*. This is analogous to the use of a D channel in basic access ISDN. So although there is a 64 kbs clear-channel capability available in PRI, the signaling system is made more complex.

The common-channel signaling for PRI is added not only to allow clear-channel transmission but to give the extra features needed to support basic rate ISDN. One feature of ISDN that has been much publicized is *calling party indication*. This features informs the person being called of the identity (for instance, the telephone number) of the person calling from the other end. The called party can then decide whether to answer the phone. In non-ISDN networks, links between the COs are T1 links and as such only have the capability of requesting a telephone link and are not able to pass on the number of the telephone that originates the call. The D channel in PRI can be used to provide such information. The PRI interface is an integral part of the overall deployment of ISDN.

Even though the use of the D channel is similar to that of basic rate, it is not identical because the applications are not the same. For example, in a basic rate S interface there can be more than one terminal on a network. For PRI, only two stations are allowed on one link, that is, it is a point-to-point configuration. When the two systems are integrated, care must be taken to realize that some sort of translation is included.

As with the basic rate, PRI signaling is divided into layers as per the OSI model. The protocol used across the PRI is *signaling system number 7*, more commonly abbreviated as SS#7.

PRI adds features other than calling party identification, such as higher-speed data channels. The B channels on a PRI interface can be concatenated to form H channels. Two H channel formats exist, H0 and H11 formats. The H0 channel is six B channels joined together to form a 384 kbs data channel. This channel can be used for either high-speed data or digitized video. For even higher speeds the H10 channel allows rates of 1.536 Mbs. This can be used for communication between host computers over large distances.

Although already proposed, it may be some time before the H channels become widely available in the public network. Nearly all switch matrices in exchanges are based on the allocation of 64 kbs 8-bit time slots. To be able to provide H channel paths through the switch, the software controlling the exchange will have to be upgraded. Fortunately the problem is reduced by the fact that the H channels are multiples of the 8-bit 64 kbs time slots.

Initially T1 was used to link COs and consequently was controlled by the telephone companies. The prime objective of T1 links was to reduce the number of physical connections between COs and hence reduce cost and increase reliability. In the past few years, the T1 network has been finding its way into the CPE. The T1 spans were initially used for the same reasons that they were originally designed for; however, these networks are now being used to provide mixed voice and data services between locations.

A T1 link can be used by the telephone company to provide a multiple telephone link within the public network. Alternatively, the whole T1 span can be used as if it were a single telephone line between two locations. This is known as a *nailed-up link*. This link can carry whatever information the subscriber allocates to the channel. At either end of the link there will be a *T1 multiplexer* or *network multiplexer*. The multiplexer will handle the interface between the subscriber services and the T1 line. In the case of a T1 network multiplexer, the equipment will also handle routing information over different T1 spans. The subscriber channels can be at various data rates and if the rates are lower than 64 kbs more than one subscriber channel can be multiplexed on one 64 kbs channel.

In many of the current multiplexer applications, proprietary schemes are used to pass signaling information and to multiplex the data onto the 64 kbs channels. With ISDN, more standards are being adopted to address the voice/data applications throughout the public network. It still remains to be seen if these standards will be adopted by the multiplexer vendors.

T1 Applications within the ISDN Environment

In Chapter 5, it was seen that a majority of ISDN development has been in the area of on-premise equipment. For instance, semiconductor manufacturers have had S interface devices for a number of years. The result of this is a growth of ISDN as a series of islands rather than as a coherent global network. The lack of standards for the network as a whole is only one reason for this pattern of development. Another reason is the availability of other solutions that integrate voice and data networks.

T1 multiplexers provide a readily available method of integrating voice and data on the same network. Voice/data PABXs have been available for some time now although they use proprietary implementations to integrate the voice and data.

The emergence of ISDN in on-premise equipment will probably continue to develop in islands until equipment is made available to upgrade the connection from the PABX to the CO or until the T1 multiplexers are upgraded to handle ISDN BRIs. Fortunately both approaches are being explored. One of the factors that facilitates this development is the established architecture of digital exchanges.

The Exchange Line Card

The PCM backplane structure that was explained in chapter 5 is also applicable to a T1 environment. In digital exchanges, the PCM backplane is used to connect T1 line cards as well as other line cards. It is no surprise then that the addition of ISDN to T1 networks for the PABX is mainly an issue of adding the line card to the ISDN reference point. The amount of work that is involved in the integration step is mainly a function of which services are supported.

Many PABX/COs now have access to clear-channel capability on the T1 interface even though it is not an ISDN link. This means that the 64 kbs channels can be switched through to the T1 link. However, if the T1 line does not support the ISDN services, only a simple ISDN implementation can be realized. This type of approach would involve little change to the T1 interface. Of course, the ability to handle the clear channel would have to be added.

A further step can be taken to add PRI services to the T1 interface. This would involve changes to both the hardware and software of a line card design. These changes would be much along the same lines as those for an analog line card. Hardware and software would be added to the T1 line card to handle the common -channel signaling. This addition could be fashioned around a basic rate design. The main factor to bear in mind is that the D channel in PRI is at 64 kbs, not 16 kbs. To offset this, the D channel processing is common to the 23 B channels so each channel does not need its own individual processor.

Currently a T1 line card contains four functional elements: the line interface; frame aligner; PCM interface; and line card microprocessor. Because the microprocessor controls the layer 1 functions of the T1 line and communication with the line card controller, a relatively simple machine will suffice, for example, an 8-bit microcontroller.

The line interface performs two functions. It will handle the level translation from the TTL levels used internally on the line card to the AMI coding (see chapter 2) used on the T1 interface. The line interface is also responsible for extracting the clocking signals from the line. These signals are used to phase lock the internal timing of the PCM highway. Because the internal rate of digital exchanges is 2,048 kHz (to allow 32 channels of 64 kbs), it has to be synchronized to the T1 line, which runs at 1,544 kHz—24 channels of 64 kbs plus a framing bit. The timing signal that both systems have in common is the 8 kHz clock.

This synchronization plays an important role in exchanges, including ISDN BRIs. The timing for such interfaces is derived from the PCM highway clocks. Any standard that is used to define the performance of the ISDN line card must reflect back to become a requirement for the performance of the PCM system clocking. Jitter is a good example of this. The S interface has a jitter tolerance of 7%. The design of the line card will inevitably add jitter between the PCM highway and the line interface, so the jitter on the PCM clocks must be less than this figure. Normally this would lie in the range of 3%.

If the PCM clocking is tied to T1 line timing, then the specifications of the phase locking circuit must take this into consideration. In a standard digital exchange, the jitter specifications on the PCM highway could be less lax only when dealing with analog telephone interfaces.

While dealing with the subject of jitter and synchronization, it is worth mentioning that one of the functions of the T1 frame aligner is to provide a buffer to accommodate the short-term variations in line clock frequencies. This short-term variation is mainly due to cumulative effects through the network, because a T1 link could pass through several repeaters and exchanges before it reaches its final destination. The buffer in the transmit side fills when the exchange clock speed is less than the line clock and empties when the line clock is slower than the exchange clock. The opposite is of course true for the receive direction. Because the amount of data is constantly varying in these buffers, they are referred to as *elastic* or *slip buffers*. These buffers are still needed whether a standard T1 line card or PRI line card is constructed.

The other function of the frame aligner is to take data from the PCM interface and put it into the 24-channel T1 format. This will include adding in the signaling for the T1 line. As pointed out previously, (see p. 105) the framing bit is used to carry additional information to the framing patterns. Two examples of this data are the *facility data link* (FDL) channel and the *cyclic redundancy check* (CRC) (a full description of the framing pattern for ESF is given in Figures 6.2a and 6.2b). The FDL channel is a channel within the framing bit transmission that enables information concerning the T1 network to be exchanged between access points, or nodes, on the network. This type of connectivity is useful, for example, if a T1 span goes down within a link. The FDL channel can be used to reconfigure the path through the T1 network.

Primary Rate User–Network Interface

Interface for the 1544 kb primary rate:

Code: AMI, B8ZS
Frame length: 193 bit, Nr. 1......193
Organization: F bit (bit 1) + 24 timeslots (8 bit each)
Framing frequency: 8 kHz

Frame structure:

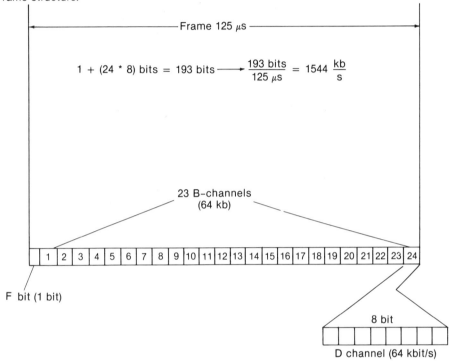

Figure 6.2a.   Primary Rate User-Network Interface.

The CRC is a direct result of the addition of data channels on the T1 link. A CRC is performed on the data in a whole 24-channel frame, including the framing bit but excluding the CRC field, for 24 frames. The CRC is then transmitted together with the other bits and the framing pattern and is checked at the receiving end. If the receiving end detects an error, it then sends an error signal back to the transmitting station. This adds an extra layer of error checking on the T1 line to reduce data errors.

The CRC and FDL channel are not specific to ISDN. They are mentioned because they do have an impact on the ISDN line card design cycle. In many cases the decision to add ISDN capability is also made at the same time as the decision to add ESF and the extra features. This entails not only adding extra hardware to handle the ISDN changes but also some circuitry for the CRC and the FDL chan-

Multiframe Structure

Multiframe frame number	F Bits			
	Multiframe bit number	Assignments		
		FAS	Note 1	Note 2
1	0	–	m	–
2	193	–	–	c_1
3	386	–	m	–
4	579	0	–	–
5	772	–	m	–
6	965	–	–	c_2
7	1158	–	m	–
8	1351	0	–	–
9	1544	–	m	–
10	1737	–	–	c_3
11	1930	–	m	–
12	2123	1	–	–
13	2316	–	m	–
14	2509	–	–	c_4
15	2702	–	m	–
16	2895	0	–	–
17	3011	–	m	–
18	3281	–	–	c_5
19	3474	–	m	–
20	3667	1	–	–
21	3860	–	m	–
22	4053	–	–	c_6
23	4246	–	m	–
24	4439	1	–	–

Note:

FAS: Frame alignment signal (....001011....).
Each multiframe is 24 frames long and is defined by the multiframe alignment signal (FAS) formed by every fourth F bit.
The bits c_1 to c_6 are used for error checking in CCITT recommendation G 704.
The use of the m bits is for further study (for example, for maintenance and operational information)

Figure 6.2b.   Multiframe Structure.

nel. Fortunately there are frame aligner devices that handle both the CRC generation and checking and split out the FDL channel. These frame aligners can be used for ISDN PRI and ESF line card designs.

The FDL channel does have its own associated protocol. This must be handled in addition to the D channel signaling protocol in a PRI interface design. The FDL protocol is a frame-based protocol, unfortunately not HDLC, so the addition of a controller could help reduce the processing power needed for the whole line card. Fortunately though, the FDL channel is quite a slow channel and requires a low processor overhead.

The opposite is true for the D channel. The protocol over this channel is running at 64 kbs. In addition, the T1 line must be maintained together with an interface to the line card controller to handle the 23 B channels from the exchange switch matrix. In many cases there is a T1 line card within the exchange that handles the standard 12-frame T1 with robbed-bit signaling. This will include the processing power to maintain the T1 line and interface to the rest of the exchange.

One approach is to leave this design intact and augment it with a second system to handle the PRI interface. To do this the current design must be able to handle common-channel signaling and turn off the bit robbing. The D channel information is taken from the interface to the PCM highway. In fact it may even be possible to locate the D channel controller remotely and switch the D channel on a PCM highway. To supplement the D channel control, an interface between the D channel control and the line maintenance/PCM interface system must be added. If the D channel control is on the line card itself, this can be implemented as a common block of memory. Given the case in which the D channel controller is situated off-card, additional interface messages must be transferred over the line card control PCM channel.

A second approach is to redesign the T1 line card completely around the existing implementation. This is particularly true if ESF features are to be added at the same time as an ISDN upgrade. Here a more powerful processor is chosen and it will handle the line maintenance function, and interface to the rest of the exchange and D channel protocol processing. This would involve the use of a reasonably powerful processor. Alternatively, as with the BRI line card, the functionality can be split such that a microcontroller can provide a coprocessing function. For instance, the processing could be divided such that a microcontroller is used to handle the line maintenance, FDL channel, and D channel layer 2 processing. A more complex microprocessor could be used to control the microcomputer and handle the other line card functions.

Irrespective of which solution is chosen, one important factor in the design is the data transfer mechanism chosen for the D channel information. The D channel data rate is 64 kbs; this means that a byte will arrive every 125–250 μs, depending on the 0 insertion. The D channel controller must process this information four times faster than the basic rate D channel. The use of DMA to handle D channel data is more attractive in a PRI implementation. This may help to reduce the amount of processing power needed for a line card.

ISDN T1 Multiplex Equipment

Multiplexers' architecture is similar to that in an exchange. Many of the issues covered on exchange line cards apply equally to the multiplexer. However, because design of the switch matrix is generally simpler, and also because the multiplexer tends to be "closer" to the subscriber application, more functionality is included in the multiplexers. These functions include expansion of the data services and *drop/insert multiplexers*.

Although it is true that the T1 interface is a point-to-point interface, a device known as a drop/insert multiplexer has been developed for T1 networks. This device allows a T1 user to daisy-chain multiplexers together and create a pseudomultipoint network. In essence the drop/insert multiplexer receives T1 data, strips off one or more channels and replaces them with subscriber information, then transmits the new data stream to the next station. In this way the channels can be peeled off from a single T1 line at different points. Currently these channels are used for voice; more commonly, low-speed data are sub-multiplexed onto one 64 kbs channel.

ISDN capability can easily be added to this type of application. As with ex-changes, T1 multiplexers are normally based around a PCM highway structure. For T1 multiplexers the clocking frequency of the highway can be 2,048 kHz, 1,544 kHz, or 1,536 kHz, depending on the implementation. An ISDN line card can be added to a T1 multiplexer as long as the multiplexer is designed to handle the internal PCM timing. This is particularly important for a drop/insert multi-plexer. As the T1 span could be passing through as many as 22 other multiplexers, attention should be paid to jitter specifications throughout the system. An inter-face to the ISDN basic rate network should meet all of the jitter specifications laid down by the governing standards. The interface circuitry will derive its system timing from the T1 line, and thus will add any jitter appearing on the line to that produced by the line card circuitry.

T1 multiplexers provide a good means of interfacing the ISDN islands of on-premise equipment. The addition of an ISDN basic rate line card can add a lot of flexibility to a multiplexer. One of the features offered by T1 multiplexers is submultiplexing of data channels. T1 offers channels of 64 kbs, and this rate is far in excess of the rate that most terminals can handle. The rates normally used are in the range of 2,400–9,600 bs. To transmit 2,400 kbs over a 64 kbs line would require rate adaption, and most of the available bandwidth of the channel would be wasted.

To make more efficient use of the 64 kbs data rate, several data channels are multiplexed onto one 64 kbs channel. One of the most popular methods of performing the multiplexing is to use a packet-switching protocol. Packets from each of the interfaces are allowed to be transmitted on the 64 kbs highway. Unfortunately, the packet protocol used on the T1 multiplexer may not be the one present on the ISDN link. So to fully handle an ISDN interface, it may be necessary to add in a protocol conversion facility in the multiplexer.

ISDN BRIs offer the ability for two 64 kbs channels to be transferred over the line. When an ISDN line card is added into a T1 multiplexer, the D channel protocol control must also be added. In fact, a full implementation of all of the lower three layers will probably be needed to control the allocation of the B channels on the line. There is a possibility that more than one protocol imple-mentation can exist until global standards are produced. To a certain extent, the equipment at the network side dictates the D channel protocol implementation. However, it may be easier to model the protocol on an existing switch specifica-tion. This would allow equipment to function either with an exchange or a multiplexer and reduce the effort of the terminal manufacturer.

H Channel Data

The PRI offers additional data capabilities not available in ISDN BRIs. These involve the use of concatenated B channels. The H0 has 6 B channels and the H11, 24 B channels. The use of these channels poses equal challenges to manufac-turers network equipment and terminals. The data rate that is available is mainly applicable to areas such as video, mainframe computer communication, and LAN interfaces.

The first problem to be overcome is that of getting data to the PRI interface. The current standards for ISDN only define two type of interface vis-à-vis basic and primary. The 384 kbs data rate for the H0 channel is too fast for basic and would waste the bandwidth of a PRI. Therefore the originating point of such a data stream would not have an ISDN interface. This would mean that to use the H0 channel, some other medium would have to be used.

To do this an interface must be incorporated into the multiplexer/exchange equipment. This interface would have to control the physical interface and the signaling from the terminal equipment. In all probability the interface would connect to a PCM highway and take up six consecutive time slots. Additional commands would have to be added between this new line card and the central line card controlling system. To handle signaling, a protocol converter would be needed to interpret signaling from the H0 data source and translate them into signaling on the PRI. This conversion could either be provided on the line card itself or as a central service inside the exchange/multiplexer.

A second problem is controlling the data flow. In an application in which the H channel is used to interface to a LAN, the continuously available bandwidth offered by the H channels may be acceptable. However, data from the LAN may be burst in high-speed data packets, say 64 kbyte packets at 10 Mbs. Some type of elastic buffer must be provided on the line card to temporarily store data between the H channel and the LAN interface. At these data speeds the use of DMA structures would be advantageous. An example of such a line card interface is shown in Figure 6.3.

One final problem is getting the data onto the line. This is particularly true for the H11 channels. In these applications, data must be transferred at a rate of 1.536 Mbs. In all cases, some method of error detection and correction will be required

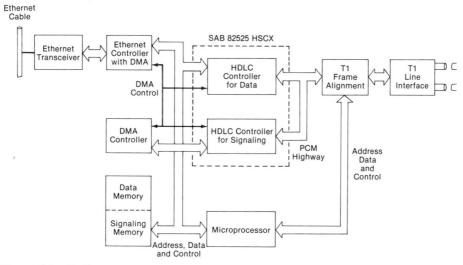

Figure 6.3.   H Channel Application: Ethernet Gateway.

for the data. One method to achieve this rate would be to use a high-speed HDLC controller. For this type of application either a dedicated microcontroller or a DMA architecture would have to be used to handle the transfer to the controller from the host system. One problem to be simplified is that of choosing the transmission medium. A modified T1 transmission scheme would be ideal for this type of application.

References

1. William Flanagan, "The Teleconnect Guide to T1 Networking," Telecom Library, Inc.

7

Software for ISDN Applications

Functional and System Description Language (SDL)

One difficulty in specifying the operation of a communications system is the variety of different applications. For example, an ISDN telephone network may be required to carry a telephone link across many different physical links. The control of the connection between the two calling parties must be maintained. This involves similar functions being performed at several points within the network. To aid in a system-independent method of defining the operation of the equipment, CCITT has evolved a nomenclature standard, Z.100, that can be used to write an operating specification in general terms.

This system is the *functional and system description language* (SDL). This language defines a set of canonical forms for different functional blocks of a communications system. These symbols break the communications system down so that its parts can be easily defined. By using this format, the different layers of the OSI model can be standardized independently of the technology used to implement them. A list of the SDL symbols is given in Figure 7.1.

The SDL nomenclature can be used to define standards throughout the seven-layer model. SDLs are used, for instance, to describe the activation and deactivation of the physical layer. At the other extreme, SDLs are used to define the networking layer operation of an S interface link.

The SDL representation is ideally suited to the type of operation that is required in a communications system. A communications system in many cases is a *state-driven environment*. This describes a system that remains in a set of given states until external stimuli cause a change within the system. In an ISDN system there can be several of these *state machines*. For instance, take the physical layer.

115

Figure 7.1. SDL Symbols

C1.5.1   *Symbols*

C1.5.1.1   *Start symbol*

It contains the process name of the process it describes (see Recommendation Z.101, § 3.3.1 and U.G. §§ D.4.3 and D.6.3).

C1.5.1.2   *State symbol*

It contains the state name or an asterisk or an asterisk followed by states names within square brackets (see Recommendation Z.101, § 3.3.1 and U.G. §§ D.4.3 and D.6.3).

C1.5.1.3   *Input symbol*

It contains the signal names separated by commas or an asterisk (see Recommendation Z.101, § 3.3.1 and U.G. §§ D.4.3 and D.6.3).

C.1.5.1.4   *Save symbol*

It contains the signal names separated by commas or a star (see Recommendation Z.101, § 3.3.1 and U.G. §§ D.4.3 and D.6.3).

C.1.5.1.5   *Decision symbol*

It contains the decision name (optional) and a formal or informal text phrase (see Recommendation Z.101, § 3.3.2 and U.G. §§ D.4.3 and D.6.3).

C.1.5.1.6   *Output symbol*

It contains the signal names separated by commas (see Recommendation Z.101, § 3.3.1 and U.G. §§ D.4.3 and D.6.3).

C.1.5.1.7   *Task symbol*

It contains the task name (optional) and a formal or informal text-phrase (see Recommendation Z.101, § 3.3.1 and U.G. §§ D.4.3 and D.6.3).

C.1.5.1.8   *Procedure call symbol*

It contains the procedure name and the actual parameters within round brackets (see Recommendation Z.103, § 2.2 and U.G. §§ D.4.3 and D.6.3).

C1.5.1.9   *Procedure start symbol*

It contains the procedure name and the formal parameters within round brackets (see Recommendation Z.103, § 2.2 and U.G. §§ D.4.3 and D.6.3).

C.1.5.1.10 *Return symbol*

It is a circle with a cross inside (see Recommendation Z.103, § 2.2 and U.G. §§ D.4.3 and D.6.3).

C.1.5.1.11 *Macro inlet symbol*

It contains the macro name (see Recommendation Z.103, § 4.2 and U.G. §§ D.4.3 and D.6.3).

C.1.5.1.12 *Macro outlet symbol*

It is a circle with a bar inside (see Recommendation Z.103, § 4.2 and U.G. §§ D.4.3 and D.6.3).

C.1.5.1.13 *Create symbol*

It contains the process name and the actual parameters within brackets (see Recommendation Z.101, § 3.3.1 and U.G. §§ D.4.3 and D.6.3).

C.1.5.1.14 *Continuous signal symbol*

It contains the condition text and then the keyword PRIORITY followed by the priority number associated (see Recommendation Z.103, § 3.3.3 and U.G. §§ D.4).

C.1.5.1.15 *Enabling condition symbol*

It contains the condition text (see U.G. §§ D.4).

C1.5.1.16 *Alternative symbol*

It contains the alternative text (see Recommendation Z.103, § 5.2 and U.G. § D.4).

C.1.5.1.17 *Join or connector symbol*

It contains the join name (see Recommendation Z.101, § 3.3.1 and U.G. § D.4).

C.1.5.1.18 *Nextstate symbols*

It contains the state name or a hyphen (see Recommendation Z.101, § 3.3.1 and U.G. § D.4).

C1.5.1.19 *Stop symbol*

It is a cross (see Recommendation Z.101, § 3.3.1 and U.G. § D.4).

In TE, this layer can be in one of eight possible states. For the physical layer the state machine is predominantly handled in hardware. Alternatively, the higher-layer state machines are more software intensive.

CCITT standards and many switch specifications use the SDL representations to define these state machines. For each state, the list of possible input stimuli is given. For each input stimulus, a series of actions is listed. The final step in the series is the end state. This can either be a different state or the same state as the start state. In this way the reaction to different input stimuli can be built up for an operating block, or process, of the protocol software. In many cases the individual states and processes are named to reflect their functions.

One example is the physical link control of the layer 2 software. In this process, the state of the physical link is maintained and used by the communications system to decide which actions are required as the call is set up and torn down. This process can be in one of several states depending on the information passed from the physical layer.

Throughout the discussion of the operation and functional description of the software in this chapter, frequent reference will be made to SDL diagrams.

Link Access Protocol D Channel

The second layer for any communications system that conforms to the OSI model is the data link layer. Because it conforms to OSI, ISDN is no exception. The data link protocol for call control is used across the D channel of ISDN and is called the link access protocol D channel (LAPD). LAPD is used throughout the ISDN to manage the transfer of higher-layer messages from different communications systems within the network.

The transfer mechanism used by LAPD is HDLC. HDLC allows data to be transmitted in packets across the D channel. In LAPD, certain restrictions are placed on the HDLC frames: first, the address field is set to two octets; second, the frame size is limited to 260 octets. The address field is split into its constituent parts as described in chapter 2. LAPD defines some special SAPIs and TEI values as follows:

TEI value 127: broadcast TEI
SAPI value 0: call control packet
SAPI value 16: packet data control
SAPI value 63: management packet

The CRC is fixed for LAPD by the polynomial defined in CCITT recommendation Q.921.

The control field of LAPD has a set of defined values for various types of frames that are used in the data link protocol. The control field also defines what type of frame is being transferred across the network. This can be one of three types:

S or supervisory frame for peer-to-peer communication
I or information frame for high-layer messages
U or unnumbered frames for additional messages

Because the LAPD protocol is part of an OSI model architecture, primitives are used to communicate between the layers. In addition to the seven OSI layers, a management layer has been defined by CCITT. This layer communicates with all layers in the system. When a service is needed from layer 2 that requires peer-to-peer communication, a request is made from the higher layer. (The management layer is considered a higher layer for this purpose.) The request will be delivered to the corresponding layer as an indication by the peer layer 2. Upon receipt of the indication, the layer will issue a response to layer 2. This is received by the peer data link and delivered as a confirm by the data link layer to the corresponding higher layer. In this way all messages from higher layers that are transferred by the LAPD are confirmed (see Figure 7.2).

LAPD is a balanced protocol. This means that the receiving end acknowledges the reception of incoming frames to the transmitting end. Only the I and certain U frames are acknowledged in this way. An S frame is used for this purpose. In the case of I frame transmission, the message from the higher layer may have to be broken down into a series of packets at the data link layer. To keep track of the order of these multiple frames, a sequence number is added in the control field of the I frame. This sequence number can either be a 3-bit number, modulo 8 counting, or a 7-bit, modulo 128, count. When the data link is set up or established across the link, the modulo counting system must also be defined.

To establish the data link, several different actions can be taken depending on the state of the processes in the data link layer. The functional operation of layer 2 is defined in CCITT standard Q.921. This standard outlines the operation of the LAPD for ISDN. Q.921 will give the different actions that are to be taken depending on the state of the various processes within the layer 2 data link control.

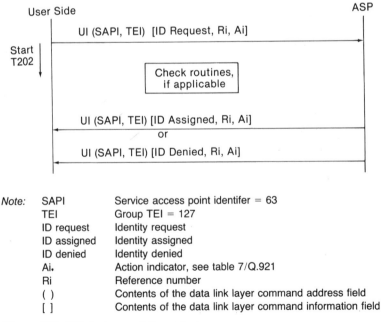

Note:

SAPI	Service access point identifer = 63
TEI	Group TEI = 127
ID request	Identity request
ID assigned	Identity assigned
ID denied	Identity denied
Ai.	Action indicator, see table 7/Q.921
Ri	Reference number
( )	Contents of the data link layer command address field
[ ]	Contents of the data link layer command information field

Figure 7.2. TEI Assignment Procedure

The process of establishing peer-to-peer communications between two data link entities can begin in one of three ways: a layer 3 primitive is issued; an activation indication is received by layer 2; or a D channel message is received. In each case the LAPD can take various actions depending on its state. For example, suppose that an establish request primitive is received from layer 3. The LAPD must check to see if the physical layer is active. If the physical layer is active, then one course of action is taken; if not, a different sequence of events will ensue. The layer 2 software must therefore keep a record of the state of the physical layer. In addition, a portion of the LAPD is responsible for controlling the physical layer. If the physical layer were activated in the previous example, then the LAPD software would check to see if a TEI had been assigned. If a TEI had not been assigned, the TE would initiate a TEI assignment procedure. The TEI, once assigned, must be stored by the LAPD software for future reference. The TE will need one, two, or sometimes three TEI values to be assigned to it. The NT must keep a record of all of the TEIs that are assigned so that no two TEIs are identical on an S interface connection.

The TEI assignment procedure can be initiated by either the NT or the TE. If the assignment procedure is initiated by the TE, a D channel message is transmitted. The transmitted message has a U frame format and is sent as an unnumbered information frame. Because there is no assigned TEI value, the broadcast TEI of 127 is used; this is a management function, so the SAPI is set to 63, that is, the management SAPI. The TE has two options when requesting a TEI value. A specific or preferred TEI value can be requested by the TE, or it can indicate that any value will be accepted. The frame format is shown in Figure 7.3.

The frame is made up of a management entity identifier, reference number, message type, and action indicator. The entity identifier will always be set to the management entity throughout the TEI assignment procedure. The reference number, Ri, is a random number from 0 to 65,535 that is used to differentiate between multiple TEs should they initiate the assignment procedure simultaneously. The message type informs the receiving entity of the category of action requested. The action indicator, Ai, is used to hold the TEI value or to signal that any TEI is acceptable.

A TEI assignment procedure would proceed as follows:

- Layer 3 issues DL-ESTABLISH-REQUEST to layer 2.
- Layer 2 issues MDL-ASSIGN-INDICATION to management entity.
- Management entity begins assignment procedure by obtaining a random number.
- Identity request UI frame is assembled and transmitted by the management entity. Timer T202 is started.
- Upon receipt of the UI identity request frame, the NT checks and assigns or denies the TEI value. If the Ai value is 127, then any available TEI is assigned.
- NT management entity transmits UI-ID assigned frame.
- Upon receipt of the ID assigned frame, the TE management entity issues a MDL-ASSIGN-REQUEST to the layer 2 data link entity.
- The layer 2 enters the TEI assigned state, stores the TEI values, stops timer T202, and continues with the DL-ESTABLISH-REQUEST procedures.

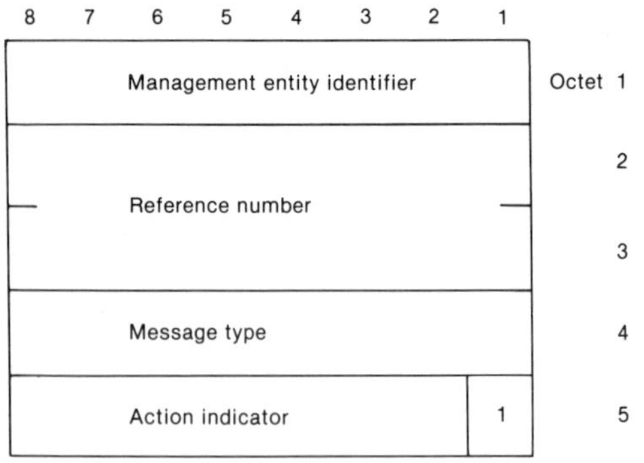

```
    8    7    6    5    4    3    2    1
  ┌──────────────────────────────────────┬─────┐
  │                                        │     │ Octet 1
  │         Management entity identifier   │     │
  ├──────────────────────────────────────┤      │
  │                                        │      │  2
  │─        Reference number            ─  │      │
  │                                        │      │  3
  ├──────────────────────────────────────┤      │
  │                                        │      │
  │            Message type                │      │  4
  ├──────────────────────────────────┬────┤      │
  │          Action indicator         │  1 │      │  5
  └──────────────────────────────────┴────┴─────┘

        Codes for message concerning TEI assignment
```

Message name	Management entity identifier	Reference number Ri	Message type	Action indicator Ai
Identity request (user to network)	0000 1111	0–65535	0000 0001	Ai = 127 = Any TEI acceptable Ai = 0–126 = Preferred TEI value
Identity assigned (network to user)	0000 1111	0–65535	0000 0010	Ai = 0–126 = Assigned TEI value
Identity denied (network to user)	0000 1111	0–65535	0000 0011	Ai = 0–127 = Denied TEI value
Identity check request (network to user)	0000 1111	Not used (coded 0)	0000 0100	Ai = 0–126 = TEI value to be checked
Identity check response (user to network)	0000 1111	0–65535	0000 0101	Ai = 0–126 = TEI value in use
Identity remove network to user)	0000 1111	Not used (coded 0)	0000 0110	Ai = 127 = Request for removal of all TEI values Ai = 0–126 = TEI value to be removed

Figure 7.3.   TEI Assignment Frame Formats.

• Different actions are taken by the management entity as various events occur—denial of TEI value or expiry of timer T202, for instance. These actions are outlined in CCITT Q.921 in Section 5.3.2.

From the above example it can be seen that several states need to be monitored and maintained, for example, TEI assignment, physical link status, and so on. The operation of LAPD software can be split into different functional blocks—one for physical link control, or one for TEI assignment control. These blocks must be able to operate independently of one another so that responses and commands that enter from the higher and lower layers can be processed. Suppose that a request from the peer entity arrived in the LAPD software while a message from layer 3 was being processed. The LAPD software must be able to respond to the incoming information as well as process the layer 3 frame. To do this a *task scheduler* is needed to realize an operating LAPD. The task scheduler will store various tasks from different software blocks, or processes, and execute them in turn. This will help ensure that all of the asynchronous tasks are completed.

As part of the task scheduler, a *queuing system* is needed for the different processes. As the tasks are originated by different processes they are entered onto the task queue of the task scheduler. The tasks can then be taken from the task queue and executed in turn. In addition to the task queue, two additional queues are required for the I and UI frames. Conditions can occur in which the frames from layer 3 cannot be immediately processed by layer 2. In these cases the frames are stored in their respective queues. The frames are then taken from the queue on a FIFO basis and transferred across the D channel to the peer data link entity. Alternatively, the incoming frames may be received faster than they are accepted by the layer 3 entity. Again a queuing mechanism is used.

In the specification of a layer 2 protocol for an exchange, the interaction of the queuing mechanism for the UI and I frames may be defined. To be able to perform the queuing mechanism effectively, the task scheduler for the LAPD software must also have some form of *memory management* associated with it. The memory management allocates the memory space needed to store the I and UI frames in their queues. When the frame has been transmitted or accepted by the higher layer, a memory release message must be provided to the memory management software to allow the memory space to be returned to the free memory area, commonly called a pool.

To complete the subsystem that is needed to set up and tear down calls, Q.931 defines the operation of the networking layer, layer 3. This layer is responsible for the following functions:

Communicating with the data link layer
Generating and interpreting layer 3 peer-to-peer messages
Managing access resources (B channels)
Checking compatibility of services provided
Call control functions, for example, timer's call reference values

Information can be sent between layer 3 entities. This can either be between a TE and NT entity (between user and network), as in the case of call set-up procedures, or from TE to TE (between user and user), as in the case of flow control for a data link across the ISDN. The frames of information used at layer 3 are composed of *information elements* that convey individual pieces of information within the message. Although the number and type of information elements in different layer 3 messages vary, each message contains three common information elements.

Protocol discriminator Layer 3 messages can be sent between user and network or between user and user. The protocol discriminator distinguishes the user/network messages from the others.

Call reference value The call reference value is assigned by the call originator. The value is only used in the user/network link and not used throughout the system.

Message type The message type defines the function of the layer 3 message that is being sent across the network.

The other information elements are added to these to form a complete frame.

Just as the layer 2 protocol had different states, the layer 3 protocol is composed of state machines with their associated states. However, in layer 3 there are substantially more individual states than in layer 2. CCITT recommendation Q.931 defines 21 states for both the user side and the network side. More states are defined in switch-specific software for different services.

The layer 3 software for a given realization is defined separately for each individual switch. The core of this specification is the CCITT specification. However, the rest of the specification is defined by the switch manufacturer. These additional specifications can either adhere to a standard or be uniquely defined. Because of this, the layer 3 specification can vary from switch to switch. This can cause substantial problems for terminal equipment designers.

General Software Requirements

Rather than describe the operation of the software needed to realize an ISDN application, an overview of the general requirements will be given. Further study of the detailed operation of the protocol software can be made by obtaining the interface specifications from the switch vendor which, in many cases, vary from switch to switch.

Protocol software for D channel control is constructed from a set of state machines. To be able to operate these machines a task scheduler or operating system is essential. This part of the software will ensure completion of all tasks generated in the execution of the protocol operation. Because the operating system can be required to perform more than one task at once, a queuing mechanism must be included. This queuing mechanism must have the ability to allocate and deallocate memory in the queues.

The operating specifications of the queuing mechanism vary depending upon the layer in which they are used. In the higher layers, the response of the memory allocation for a frame to be queued can be quite slow. However, as the incoming frames from the line arrive in a synchronous manner, the space for these frames must be allocated quite rapidly. To overcome these differing constraints, it may be better to have two separate memory allocation routines to handle the two extremes. In many cases the protocol software will part of an overall system—a voice/data workstation for example. The higher-layer memory allocation can be handled by the operating system of the equipment. There is still much to be gained by having a separate *memory management unit* (MMU) to handle the incoming frames from the line. This MMU can be relatively simple because the incoming frames can be copied to a memory area allocated by the higher-level MMU between each frame reception. If this type of scheme is adopted, then a buffer release mechanism must be added between the two MMUs.

The protocol software needs a set of timers to be able to time the various events at the different layers. The number of timers that a particular system requires will of course vary; it will commonly be in the range of 10 to 20. These can be provided as part of the operating system or as a separate software module within the completed system.

Although the queuing mechanism, MMU, and timer management are not defined in the standards or switch specifications, they are still needed. This is also true of ancillary functions that may be needed as part of a complete system. Networking statistics are just one example of this type of function. So additional work must be completed by the software designer before a system can be completely specified.

The differences between the switch specifications pose a problem to the terminal equipment software designer. There are a couple of approaches that can be adopted to solve this. One is to download the protocol software into the memory of the equipment. This can be a viable solution for a plug-in card for a PC or even a dedicated voice/data workstation. The protocol software can be loaded at power up from a file stored in nonvolatile memory (for example, on a hard disk). For a simple telephone set the problem is harder to solve. The protocol software for several switches can be stored in PROM and the correct one selected by a hardware switching option. Alternatively, the protocol software can be stored in erasable PROM and reprogrammed when the telephone is connected to a different switch. In either case the problem will be a trade-off between convenience and cost.

The OSI model is used throughout protocol specifications. However, additional partitioning is needed to design a practical system. The software that is used at the higher levels of a communications system runs asynchronously, that is, the sequential operation of the program does not follow a periodic pattern. In many cases the operation of the higher layers is controlled by the user interface. Alternatively, the lower levels of software are more synchronous in nature. The lower levels interface to the transmission media, which for ISDN means a synchronous communications system. For instance, the voice information must be transmitted every 125 μs.

To allow both of these two different blocks to function together, the communications system can be partitioned into two parts—the hardware drivers and the protocol software. This partitioning is done in parallel with the OSI architecture. As more complex integrated circuits become available, the higher-level functions are becoming part of the hardware and are not realized as a software subroutine. Many applications use data link controllers in silicon. This leads to a substantial amount of the lower three layers being implemented in hardware.

Integrated circuits in ISDN systems use, for the most part, an *interrupt-driven architecture*. In this way the integrated circuit controls the software operation to ensure that the program will service the device within the time constraints of the communications system's synchronous operation. For example, the D channel information on a BRI is transferred at the rate of 16 kbs. The data will arrive at the receiving station at the maximum rate of one byte every 500 μs. The hardware must receive and store the data into local memory. This constraint must be met for any piece of ISDN equipment. The microprocessor system must be able to react to the reception of the byte within 500 μs. This then determines the *interrupt latency time* for this operation. The program has 500 μs to stop processing the current operation, recognize the interrupt source, and store the data.

HDLC controllers are available that contain internal memory in the FIFO form. The FIFO helps to relieve the processor by increasing the interrupt latency time. A 64 byte FIFO would increase the latency time by a minimum of 16 ms. The time would be greater than this on average because of the 0 insertion that is used in the HDLC protocol. One important feature of the FIFO is its ability to store more than one frame. In certain cases the flag delimiting of the frame boundaries can be shared between frames. If this is the case, then either the processor must be able to react to this and empty the FIFO, readying it for the next frame, or the FIFO control must be able to mark the end of the frame before storing the next frame.

If the FIFO cannot store more than one frame, then the processor must be able to acknowledge the end of the frame, allocate memory space for the frame in local memory, and transfer the data from the FIFO in less than 500 μs. The advantage of using the FIFO storage to increase the interrupt latency can be lost when a shared flag is used. Care must be taken when calculating the latency times needed for HDLC frame reception.

Once it has received or transmitted a frame, the hardware driver will have to notify the higher-level software entity of the completion of the operation. This may not be immediately achievable. As pointed out earlier (page 122), the protocol software is executed on a task-scheduling basis, so the software entity that the hardware driver is trying to communicate with may not be active. To solve this problem there must be a method of temporarily storing the communications between the higher-level software entities and hardware drivers.

This can be done by using a *mailbox system*. Mailboxes, as the name suggests, allow information to be stored in temporary memory spaces until the reception entity is ready for them. Communication between the two parts of the program is achieved by the use of short sequences of bytes, called messages, that are loaded and read via the mailbox. These messages are loaded from the source process and read by the destination process. To control the flow of information through the mailbox, a signaling or *semaphore byte system* is used. This can either be in the form of a bit or byte encoded when a message is loaded or read, or as an interrupt in the task scheduler to signal that a new message is available in the mailbox.

By using mailboxes, the synchronous hardware drivers and the asynchronous higher-level protocol software can run differently and yet stay in communication with each other. In fact, the use of mailboxes is so common that it is standard to label messages that are transferred from the higher layers to the lower layers as commands and the messages in the reverse direction as responses.

The use of mailboxes solves another difficulty in realizing a practical software architecture—that of interlayer interaction. In the D channel call set-up software, the call set-up is initiated by lifting a handset. In the system lifting the handset, off-hook condition normally leads to generation of a hardware interrupt. This hardware interrupt is translated into a connect request to the layer 3 entity. In essence, the message must be routed directly from the layer 1 function to layer 3. This is not allowed in the OSI model, as layers can only communicate with the layer above and the one below. By using a mailbox, the message can be routed from layer 1 to layer 3. Other instances of this requirement occur in different areas of the software.

The mailbox structure can be accomplished in different ways, but one of the more common is to have a number of mailboxes within the system for communication between the various entities. These mailboxes can connect entities of different layers together across boundaries that span one or more layers. The mailboxes can also be used to allow communication from a layer into the management entity. In this way a complete system can be built.

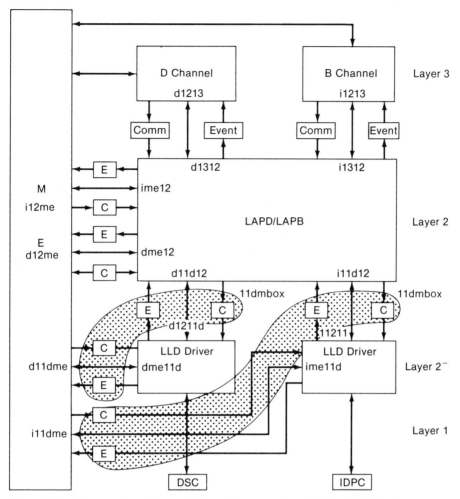

Figure 7.4.   Command and Event Mailbox Structures for AMD AmLink® Software.

When the protocol software is integrated into an existing operating system, as in the example of a voice/data workstation in which the data portion may already exist, mailboxes can be used to link the two sets of programs. An existing operating system can be used to run the protocol software. The user interface would send commands via the operating system to the protocol software using one mailbox. The protocol software would in turn request services, such as memory allocation, from the operating system through a different mailbox (see Figure 7.4).

Device Driver Software

The lower functions of a communications system's software must react with the circuitry that interfaces to the physical medium. The portion of the software that is responsible for this task is the *device driver software*. The devices themselves

are becoming more complex, so the device driver software is increasingly becoming a larger part of the overall system. More functionality is needed for this
functional block.

In many cases the integrated circuits used in telecommunications devices have
several programmable registers. These registers fall into five broad categories:

Configuration registers
Status registers
Command registers
Interrupt registers
Data registers

The configuration registers are increasing in number as integrated circuits
become more complex. The goal of the IC designer is to try to address as many
applications as possible. To do this, as much application-specific circuitry as
possible is included on the chip, along with a set of configuration registers to
tailor the circuitry to an end-user application. The problem of designing hardware
for communications equipment is as much one of programming the right configuration of the devices as picking the chips in the first place.

The hardware and software for communications equipment must often satisfy
different operating environments. To do this a set of *configuration tables* is
included in the device driver software. These tables are used to configure the
hardware dynamically from one application to the next. This approach is not new;
the difference in ISDN designs is the variety of configurations. The same hardware
design may be used for different applications depending on the configuration of
the ICs. One example of this approach is the data service of an ISDN terminal. The
user of the terminal could decide to use one of three different types of data
transfer protocol: X.25, V.110, or V.120. The only difference between the data
services is the software. Once the initialization of the hardware has been accomplished by the device drivers, the appropriate protocol software can be executed.

Another use for embedded tables in the device driver software is for set
patterns of data. These set patterns could range from test patterns for diagnostic
purposes, to D channel frames of data, to digitized voice information. The last
example could be used to provide call progress information in a terminal either as
a set of tones or as digitized speech.

The programming that is used in the device driver software must be able to
execute quickly enough to react to conditions on the physical interfaces. Some
form of management entity must be included in this block. For example, the
device driver would have to react to a loss of line synchronization. The loss
would be detected by the line interface device and reported to the device driver
software as an interrupt. The device driver software would have to react to this
interrupt in real time. For instance, it may take too long for the task scheduler to
allot time for the error processing software to handle this error condition. Other
errors could be handled by a central error processing module. A decision must be
made on which errors can be handled within a task scheduling environment and
which are to be handled by the device driver.

Part of the error processing in the ICs is *interrupt generation*. In some of the
more complex devices the number of interrupt sources can run into a consider-

able number. The problem with multiple interrupt sources is how to decide the priority of each interrupt. The priority determines which interrupt is processed first if several are reported at one time. There is no one set of interrupt priority ordering for a given device; the order must be decided by the device driver designer and the system software designer. The impact of each interrupt on the system must be evaluated; its relative priority can then be determined. In certain cases one event may cause several interrupt conditions to be reported. For example, take the loss of synchronization during the reception of the CRC of a D channel frame. The following interrupt conditions would be reported by the hardware: loss of synchronization; receiver CRC bad; and incomplete received frame.

The device driver software may choose to ignore the latter two interrupt conditions. The processing of the interrupts is a decision of the device driver designer. However, the designer will need to discuss the interrupt processing with those who designed the overall system, the protocol software, and the hardware.

In certain instances there may be interaction between the interrupt service routines and the higher-level functions. This would occur when an incoming frame is received on the physical interface. The device driver software would be notified by an interrupt, and as part of the interrupt service routine, memory must be allocated by the operating system. Depending on the hardware chosen, this could demand a very rapid routine to be able to meet the interrupt latency. In addition, in a data application it is possible to receive three such requests in one 125 μs time frame if both B channels are used for packetized data transfers. It may be easier to design a MMU for the HDLC device drivers independently of the operating system MMU. Each interrupt service routine would have to keep a record of the memory allocated for the incoming frames.

Most of the CCITT line interface specifications include maintenance functions. These functions include loopbacks and self-test requirements. These requirements, together with power-up self-test, lead to a test block being part of the device driver software. Some of this burden can be off-loaded onto hardware by using special test facilities in the ICs themselves, special configurations, or by adding additional circuitry for these functions. Although the specifications call for these functions, not all of them are completely defined (self-test, for instance). The hardware designer and the device driver designer must decide what constitutes a complete self-test.

Some other special requirements may exist for specific line interfaces or applications. The storage of the TEI value is a good case in point. This value must be stored when the terminal equipment goes into a power-down mode of operation when an emergency power condition is detected on the line (see recommendation I.430). A special memory area may have to be allocated for this type of data.

8
Testing ISDN Systems

Introduction

As with all other aspects of ISDN, the area of testing ISDN equipment brings new challenges with it. The revolutionary technology of ISDN forces test equipment manufacturers to address these new challenges when they develop new products. The new test equipment can be upgraded to test existing products, or must be defined and manufactured anew.

The existing analog telephone network has been in use for a number of years; consequently, there is a wealth of data on the optimum performance criteria for a given operating environment. This type of data takes time to collect. To collect the same type of data for ISDN equipment may not take as much time as it did for analog telephones, but it will take some time. Until ISDN becomes a major portion of telephone networks, the standards and test specifications will be changed accordingly.

The complexity of ISDN equipment causes difficulties in the testing phase of a design project. To be able to effectively test an ISDN design completely requires a great deal of expensive equipment. This equipment can either be purchased by the designer, or a fully equipped laboratory must be leased. The cost and amount of use a piece of test equipment may get will probably be the factors that decide whether to lease or buy the equipment. In certain cases the test *equipment* that is required may be too expensive to either lease or buy, for example, as in the case of an ISDN exchange. In this case, test *facilities* must be leased.

Although the CCITT standards work well in terms of defining the mandatory operating parameters for ISDN terminals and line cards, they do not specify the individual components that go to make up the system. For instance, the line interface consists of both an integrated circuit and the associated transformers. Each of these components must be tested separately before they are assembled in a circuit. However, the component test must ensure that when the devices are connected together they will meet the CCITT standards. A set of parametric limits must be defined such that the rest of the system is simulated at the device's test points. So in this case not only does the testing strategy have to be defined, but also the test limits and loading conditions.

Physical Interface (Layer 1) Testing

Before designing a test plan for the physical layer, it important to define exactly what needs to be tested at this layer. The physical medium can play an important role in determining the overall performance a system. A high bit error rate due to a failure to meet a physical interface parameter will affect the software performance in layers 2 through 7 to one degree or another. Physical layer performance statistics must also be characterized to be able to optimize the system operation in general.

One of the major problems of analyzing the physical layer is that it is almost mandatory to have a completely functional ISDN environment to fully evaluate the impact of parametric variances. This may be possible for a sophisticated lab set-up, but is highly undesirable for a production test. Many basic parametric tests (on a terminal, for example) cannot be carried out without a network termination of some sort. A production test system for a terminal must therefore either include a network termination or be able to generate the signals synthetically.

Having to provide a mixed digital/analog test system for ISDN production testing also poses problems for manufacturers of ISDN equipment. Although the signals on the telephone line are digital in terms of the information that they represent, they are still analog pulses that have to measured using analog means. The output pulses from the equipment must be measured—however simple the measurement—to be able to guarantee that the system will meet the CCITT standards. For instance, it may not be necessary to perform a full pulse mask template measurement on every system that leaves the production area.

Although more and more test equipment is becoming available as ISDN gains ground, there are still many gaps in the product offerings. Test equipment manufacturers labor under the same, basic difficulty as manufacturers of ISDN systems: which areas of the test cycle will provide the best market potential? Also, with more and more test equipment becoming computer controlled, the ability to build a custom test system out of standard, off-the-shelf products is increasing.

The greatest boon to providing this ease of test equipment integration is the general acceptance of the IEEE-488 interface. Almost any piece of standard lab equipment can be obtained with an IEEE-488 bus interface. Many packages are on the market that turn the ubiquitous PC into an IEEE-488 controller. Some of these systems are rudimentary display programs; others are full-blown control and analysis packages. More and more test equipment manufacturers are adding IEEE-488 busses to their product offerings and IEEE-488 PC controller packages to their product lines.

Perhaps the single most essential piece of test equipment needed for layer 1 testing is a digitizer. This can either be in the form of a digital oscilloscope or as an add-on to an existing tester. In a development environment, a digitizing oscilloscope is essential. With the increasing complexity of these instruments it is possible either to perform many of the line interface measurements with the oscilloscope alone or to enhance the capabilities by using computer control.

Pulse template measurement is one area in which the use of computer-aided instrumentation is advantageous. During the development cycle, the pulse measurement may be made to verify the performance of the final design. A pulse template measurement is required to verify the transformer, IC, and protection circuitry in the initial stages to ensure no extraneous effects are introduced in the layout. The output pulse could be captured by the digital oscilloscope and then be down loaded to a PC and superimposed onto the I.430 template for analysis.

These measurements could be stored on the PC disk for future reference. In addition to controlling the digital oscilloscope, a PC can operate other pieces of test equipment. For example, an IEEE-488–compatible power supply could be added to the test environment. Under the control of a PC, the power supply of a piece of ISDN equipment could be varied and a series of parametric measurements, for example, pulse templates, could be recorded. This would allow correlation of the variation of one parameter with respect to another.

By constructing a test bed that is centered around a computer-controlled system, some extremely complex measurements can be made. The previous example could be enhanced to allow the study of the interaction of several different parameters. This could be accomplished by using the computer to control the stepping of the parameter variations. As each change is made, several parameters can be measured and the values recorded. The changes could be performed in such a way that a full set of readings could be made for the variations of the different control parameters.

To show this type of measurement system in action, suppose a correlation of supply variation, white noise on the line, line length, and bit error rates were to be made. The computer could step the voltage from 4.5 V to 5.5 V in 0.1 V increments; the noise level from −80 dB to −10 dB in 5 dB increments; and the line length from 0 km to 1 km in 100 m increments. The algorithm in figure 8.1 could be used to control the system and measure the bit error rates.

This algorithm assumes that all of the equipment used can be computer controlled. At the end of the measurement process a data base of readings would be created. Different sections could be taken through the data base to create graphic representations. For instance, a histogram could be made showing line length versus bit error rates, and using different colored bars for different noise levels on the line. This graph would have to state the power supply voltage that was used for that particular series of readings (see Figure 8.2).

By using a computer-controlled system, a large amount of performance data can be collected in a lab setting that would allow a designer to predict the performance of ISDN equipment in a system. Another advantage of using computer

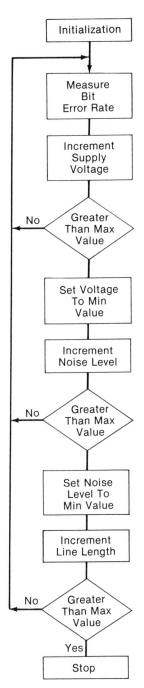

Figure 8.1.   Flow Chart for
Line Measurement Example.

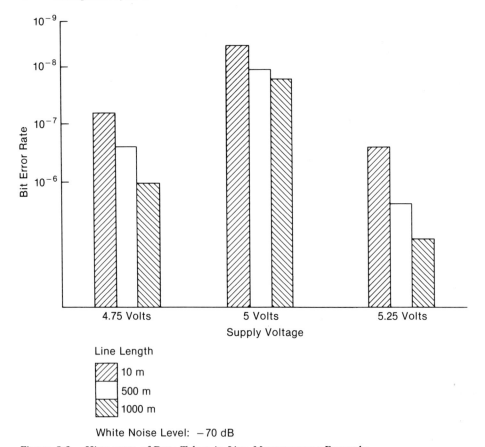

Figure 8.2.   Histogram of Data Taken in Line Measurement Example.

-based measurement systems is the ability of a PC to monitor for a particular event over long periods of time. The analysis of sporadic anomalies in the behavior of an ISDN system can require a great deal of time. The major problem is capturing that single, rare event that causes a problem. In some cases it may even be necessary to use special features of more modern types of test equipment.

A piece of test equipment that would fall into this category is a logic analyzer combined with a digital oscilloscope. The following scenario illustrates how this type of instrument could be used.

A TE periodically and briefly causes the telephone line to stop passing voice information. This could be caused by many different failure mechanisms in an ISDN system, but one could be the loss of synchronization on the line interface. The logic analyzer is set to activate when it reads the particular status register that indicates a loss of synchronization. The digital oscilloscope in the instrument measures the activity on the telephone earpeice and the line receiver at the same time. When the logic analyzer captures a loss of synchronization, measurements of the line and audio activity are made. A determination of how long the line is inactive and whether this could be detected by a user can be made.

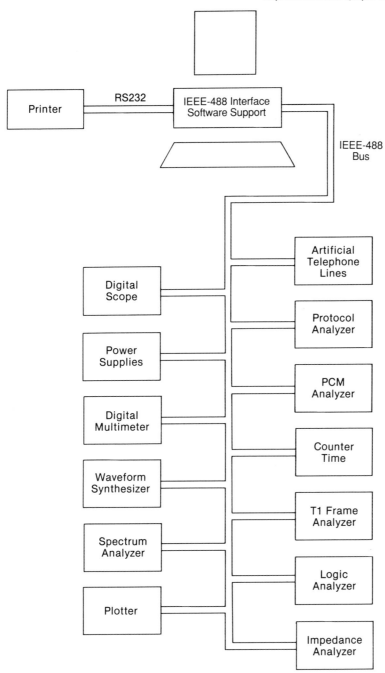

Figure 8.3.   Test Bed for ISDN.

By carefully choosing the test instrumentation for a lab set-up, many system-oriented measurements can be made in the laboratory. For ISDN this may be the only way that information concerning equipment performance can be obtained. This would give a wealth of performance data for ISDN equipment that may even be equal to that available for analog equipment, but in a very short time frame.

When making any measurements on the line it is essential to have a piece of equipment to simulate the rest of the ISDN network. For layer 1 measurements this may simply involve a device that allows the completion of an activation sequence and source and sink 2B+D data. For more complex tasks a more sophisticated piece of equipment may be needed.

The designer may find that not only is the task of making the final product demanding, but also that designing the test environment requires considerable effort. Here again a PC-controlled environment can help solve the problem. The PC can be used to control both the measurement system and the simulation of the rest of the ISDN network.

Finally, a key piece of equipment in any layer 1 test system is a printer/plotter. This device can produce hard copy output of any measurements that are made. These copies can be used to help solve design problem or as part of the performance verification of the final product. By using a PC-based system, the addition of both a printer and plotter is possible. The printer is used to output information such as tables of readings, whereas the plotter can be used to provide graphic representation of the data and waveform diagrams. A typical system of a layer 1 test bed is given in Figure 8.3.

Layer 2 and 3 Testing

A large portion of any ISDN design is the software development; a large part of that is the software testing. In general the amount of time required to adequately test a software design can easily be underestimated. This is especially true for ISDN designs in which there may be little or no previous experience of communications software development on which to base a realistic decision. The terminal end of the ISDN network is probably the area where an underestimation is most likely.

In terminal equipment design there is a good chance that the equipment designer has little experience with communications software development. This can be due either to the terminal design being an add-on card for a computer system or an ISDN telephone. It is not immediately obvious that the design of either of these products would entail considerable software development overhead. Additionally, because the developers of ICs are incorporating more and more circuitry onto the devices, the amount of effort in the hardware design side is reduced. This does not significantly affect the software development in the same way. The testing of a hardware design may take about 10–20% of that needed for the software.

Another factor to be considered when putting together a software testing plan is the equipment required. For instance, what is needed to be able to test a call set-up and tear-down both in a development and production environment? The equipment involved can easily run into the hundreds of thousands of dollars. Obviously this will be a significant factor in budgeting the cost of development for an ISDN product.

The software testing can be broken down into three phases: static testing; integration and dynamic testing; and system testing. In each phase the amount and type of test equipment will vary. In the latter phase there may even be a *conformance test* to pass before the product can be connected to the ISDN network.

As detailed in Chapter 7, much of the software used in ISDN equipment is state driven. This type of structure lends itself to systematic *static testing*. A static test involves single stepping the software through its operation to verify that it performs according to the SDL representation in the specification. To do this a special version of the operating system interface can be developed such that the system halts after the execution of each step. Stimuli can be input into the system either by an external piece of hardware specifically designed for the task, or under software control.

The latter case is more desirable from the standpoint of including special testing that is specific to a particular design. Furthermore, if hardware responses can be input under software control, then the possibility exists to develop the higher-layer software without the hardware. This can be a considerable asset in reducing the development time of an ISDN project as software development can begin even before the hardware system design is complete.

Although there are debug tools available to test software at a high level, that is, at C source level, it may be advisable to investigate a tooling system that allows a testing philosophy that correlates to the SDL representation. Just as a C source code debugger can step through modules, subroutines, and even single lines of code, a test tool with a higher level of granularity at the SDL level can speed up software testing. This type of test tool would allow the test engineer to step through the code at an SDL representation level. The test scripts for the system could also be defined at this level.

An important part of the static test cycle is documentation of the testing. This documentation should relate both to the source code and to the layer 2 and 3 software specifications. In terminal equipment designs, a software specification is provided by the exchange manufacturer. If possible, a specification of the conformance testing, which is mandatory before equipment can be connected to the network, should be obtained. The software testing and documentation should conform to this specification.

After the higher-level software has been static tested, the next step is to integrate the code into the system and perform *dynamic testing*. Dynamic testing, as the name implies, involves monitoring the software operation in an active environment. This can either be a real ISDN network or a simulation. The major problem with this stage is cost. If the software is to be tested in a real system, the equipment must be obtained to realize such a system. In the case of line card software testing, this may not be a great problem if a terminal exists for the switch specification. However, if the line card software is to be fully tested, then each line must be downloaded. Testing the terminal equipment in this fashion would require an ISDN exchange, not a exactly a cheap piece of test equipment. In some cases though, this type of testing can be desirable, particularly in the case of a manufacturer developing both the terminal equipment and the exchange.

If a real system is not available, then some form of simulation must be found. ISDN simulators are becoming more common. Still, the cost of these systems can range from a few thousand dollars to tens of thousands of dollars, depending on the sophistication of the system. In some cases the approach that is taken is to develop the "other end" of the network from the equipment that is being developed. This method does have two drawbacks. The first is that the amount of development needed may at best be equal to that required for the actual product. The second is that software errors can be replicated as a "mirror image" in the test software.

The use of simulators to perform dynamic testing does offer the advantage that the testing cycle can be accomplished much more easily. Products are available, such as the CPE test bed from Tekelec (see figure 8.4), that allow a software designer to have the equivalent of the rest of the ISDN network in a lab environment. In addition this system also works in a monitor mode that is extremely useful in the system testing.

A step down from a complete package to simulate the ISDN network is a *protocol analyzer*. Many equipment manufacturers produce protocol analyzers

---

ISDN CPE Testbed™

Applications software for the
Chameleon® 32

Only Tekelec's Chameleon® 32 Gives You Affordable ISDN CPE Conformance Testing Today

As a CPE vendor or operating company, you're faced with a variety of constantly changing ISDN standards in the U.S. market. To ensure your equipment's compatibility with switches from various manufacturers, you need a test system that is flexible enough to keep up with changing standards and is backed by an industry leader.

That's why Tekelec offers the ISDN CPE Testbed software package, incorporating Bellcore-licensed software, for the Chameleon 32 ISDN Test System. We've taken Bellcore's exploratory technology and turned it into a commercial product that can save you valuable development time -- and money.

Flexible Software Tests Compatibility With Major Switch Manufacturer Standards

The pre-written package lets you verify your products' compatibility to the Bellcore TR-TSY-000268™ specification, as well as each of the five major ISDN switch standards, which include:

- AT&T 5ESS™
- Northern Telecom's DMS-100™
- Siemen's EWSD™
- NEC's NEAX61E™
- Ericsson's AXE™

A unique NT emulator simulates the NT side of an ISDN interface and allows you to monitor your equipment's responses. Tests may be run over an ISDN Basic (2B + D) or DS1 Primary (23B + D) Rate interface.

Extensive Tests for Levels 2 and 3 Ensure CPE Compliance

The ISDN CPE Testbed contains a comprehensive series of data link and network layer tests. An easy-to-use test script language lets you send specific layer two and three messages to the CPE, as well as define the CPE's expected responses. Test capabilities include:

- Q.921 and Q.931 protocol tests
- Circuit- and packet- switched call control
- B-channel continuity
- Error-handling procedures
- Supplementary services for both voice and data
- Service interactions

In addition, a TE emulator provides a powerful tool for developing and debugging additional user-written test scripts.

Eliminate Unnecessary Development Costs

The Chameleon 32's pre-written ISDN CPE tests save you valuable development dollars. You can repeat the tests as needed, against as many new products and product releases as you wish. Having this powerful package in-house means you don't have to pay for repeated trips to a certification lab. And because the switch manufacturers are constantly updating their standards, Tekelec offers an optional 12 or 24 month software maintenance and update program so you can keep up with these changes as they occur.

Make Your Commitment to ISDN Compatibility Now

Call 1-800-TEKELEC today to schedule your free demonstration of the Chameleon 32's ISDN CPE Tests.

Figure 8.4.   ISDN CPE Testbed™ from Tekelec.

for other data and communications networks, for example, X.25 SNA. These pieces of equipment can be modified by adding a line interface and the necessary software to perform the same function for ISDN. In almost all cases some form of programming language—C, FORTH, or BASIC are the most popular—allows the user to define a test program to simulate the system. These pieces of equipment can also be used in system testing. When fitted with some form of mass storage (hard or floppy disk), they can be indispensable tools for system debugging.

When static and dynamic testing have been completed, the prototype is ready to be tested in a real system. This may involve a conformance test for terminal equipment in which the manufacturer ensures that the operation of the new equipment will not cause any malfunctions on a live switch. The system test is probably the most difficult to perform. Errors may happen sporadically over long time intervals. This type of error may be difficult to recreate for analysis in a controlled environment. The use of test equipment such as a protocol analyzer can help to track down this type of niggling problem.

The addition of a *log file generation utility* in the operating software of the ISDN equipment can be of invaluable assistance in the system debugging phase. These files can be generated over long periods of time and can capture a particular event that needs further analysis. The inclusion of this type of utility must be decided upon at the system specification stage during software development.

Self-Tests

The previous section outlined external tests that are performed during the design and development of an ISDN product. When the final product is in use there still remains the self-testing that is performed by the equipment itself. Certain self-tests are outlined in the communications standards but these tests mainly involve loopbacks at various points in the network. The ability to set these loopbacks allows the self-testing to include the remote station.

A maintenance channel sets a system's remote loopbacks. As terminals and line cards increase in their ability to perform complex functions due to the processing power available, more complex remote loopback testing can be utilized. The main use for remote testing is to verify the operation of the transmission media. A loopback is set to return all of the data that are transmitted from the originating station. The data stream can then be analyzed for bit errors, and so on.

Even though these standards exist, they can only work if the test programs are embedded in the equipment. Part of the final software should include these types of testing. This allows the equipment not only to detect a failure in operation but automatically go some way in beginning to discover the possible cause. The use of remote and local loopbacks to accomplish this is becoming more widespread.

Another type of remote testing that is appearing in the maintenance channel specifications is the request for the remote end to perform a self-test. Part of this self-test procedure should include some form local loopback testing. In this way the various access points to the network can be tested.

Because even a simple telephone will need some form of microprocessor system, a general system test routine will be incorporated as part of the final software. This would also most likely apply to line card architectures. These tests would include such things as peripheral testing and memory testing. This type of testing is performed as a matter of routine in the computer industry so finding examples of it will not be difficult. In fact if an off-the-shelf operating system is chosen, self-testing and diagnostic routines can be acquired quite easily.

Certain test procedures have already been defined for layer 1. These functions mainly involve the setting of local or remote loopbacks on the line. The remote loopbacks can be set from either end of the line. The station requesting a remote loopback will send a loopback request, and the other station will send an acknowledgment when a loopback has been established. In addition to the loopbacks, other test functions can be requested. These test functions can either be a request for a self-test or a bad CRC to test the layer 2 functions. These requests and acknowledgments are sent over the S/Q channel in the S interface and the *embedded operations channel* (EOC) in the U interface. In both, this auxiliary information is sent over a subchannel that makes up part of the layer 1 frame. Also, the bits in the maintenance channels are collected over several frames on the line in both line interfaces. The complete set of frames that contains one frame of maintenance data on the line is called a superframe. Because the collection of these bits within a superframe structure would be an arduous task for a microprocessor, it is advantageous to have an LIU that will decode and frame the maintenance channel bits.

User Boards

Many device manufacturers offer user boards to support design activities. These boards are targeted at two specific areas—evaluation and development—that are at the opposite ends of the spectrum. Evaluation involves the requirement that a hardware designer be able to study the operation of a device easily in a simulated environment. Development calls for tools that will let a hardware/software engineer build and test the final product design.

Most user boards that are provided for this purpose are PC based. This facilitates the problem of providing a user interface for the board. One additional advantage is that packages already available for the PC can be utilized. These can either be instrument control packages that aid hardware design, or software utilities for the software engineer. Furthermore, with a PC-based user board, a better technical link can be forged between the device manufacturer and the designer. Problems can be replicated on the user board system and analyzed by the device manufacturer.

If the board is to be used for software development, the inclusion of a microprocessor on it is essential. With a processor on the board, programs can be downloaded and executed in real time while the PC is used for control and analysis purposes.

In many cases, the device manufacturer will provide information on the user board design that will assist in product development. This can be in the form of schematic or layout information for hardware development, or device driver software for inclusion in the product software.

User boards can also be included as part of the test equipment. These can be fit into several different points during product development. The hardware test is the most obvious need that a user board can fulfill. A user board can be used to simulate the hardware that will exist at the opposite end of the ISDN network. This can be extrapolated so that the PC and the user board develop test programs for the layer 2 and 3 software testing. Finally, a user board can be integrated into a production test environment. This may solve the problem of finding equipment to test product-specific parameters.

9
Putting It All Together

Getting Started

Once the decision has been made to "get into ISDN," where is the best place to start? The first step is learn about the environment at which the product is aimed. There is no easy substitute for taking the time to become educated in ISDN technology. There are many ways to do this. Reading publications on the subject is one way. As ISDN becomes more entrenched in the communications world, more literature will be widely available. Another method is to participate in one or more of the many courses and seminars that are advertised. A word of caution: when looking at these courses, make sure that the course meets the requirements. It is of little use to send a design engineer to a course on cable installation. A short cut is to hire consulting organizations to aid in development. This will aid product production but can become a nightmare if not fully thought through. A product could be produced that no one in the company really understands or knows how to support.

The first step in any product cycle is definition. In ISDN, this is not as easy as it sounds. Suppose the decision has been made to build a voice/data workstation: what problems will be faced? The first, is what type of workstation should be built? Will it be a completely new design or will it be an enhancement of an existing terminal? Will the workstation be the sole data source, or can other terminals be connected to the line interface through the product? Once these types of decisions have been made, the next round concerns which standards are to be used. This not only involves the call set-up procedures defined by different switch manufacturers, but also which data transfer protocols are supported.

Because ISDN is a relatively new technology, the possibility of spinning off a

142

line of products from the initial design exists. For example, in the voice/data workstation example, once the design has been completed, the following products could be developed quite easily: standard telephone; ISDN modem; ISDN cluster controller; and plug-in cards. The decision to explore other markets or just stick to a niche area can affect the software and hardware development of the initial product.

When looking at an ISDN product idea, a careful analysis of existing solutions must be undertaken. For instance, in an ISDN voice/data workstation, how is the telephone function to be realized? There are two options available: design a special telephone set to be integrated into the overall design, or build an interface to an existing telephone into the product. Additional steps should be taken to integrate the ISDN product into existing communications system. This includes both the analog telephone network and LAN solutions. ISDN can act as a cohesive force in the communications environment as well as a complete system in its own right.

Realizing a Product

Once a product strategy has been defined, the next step is to build the product. ISDN products require careful planning in the initial stages to ensure that they are brought to the marketplace in the easiest fashion.

The first step in this phase is to put together a project team. This may sound like a simple step until the project engineer has to convince a high-level management executive that an ISDN telephone will take about 4 man-years to develop. The team should take into consideration the requirements of the product. At minimum the team will probably consist of a hardware engineer and two software engineers (one for high-level software and one for device driver software). In most cases this will develop into a team of around seven: two hardware engineers, four software engineers, and a project manager. The two hardware engineers will be responsible for the hardware design and the test set-up. The software team can be split into device driver designer, operating system designer, layer 2 designer, and layer 3 designer.

One of the first tasks that the team should address is the testing requirements. The reason that this should come first is that additional resources may well be needed to design special test equipment. The last thing that should happen is that the product be ready to go into production without a test plan ready. The differences with ISDN designs are that it is essential to simulate a network at both the development and production phases, and that substantial time and effort may have to be devoted to special test equipment. In some cases it may be advisable to involve a test engineer from day one.

A good example of how this type of problem can affect a design is to go back to the voice/data workstation and consider the testing problems there.

NT hardware is needed to test the S interface.
An exchange simulation is needed to test the layer 2 and 3 software.
An ISDN link is needed to test the voice and data connection.

A test fixture is needed in a production test to allow all three of the above to be carried out.

The equipment needed to fulfill these requirements may be as time consuming to acquire or develop as the product itself.

The next step is to start the product definitions. This can be split into two parts—hardware and software. Nevertheless, close cooperation between these groups should be maintained as decisions made by one group can drastically affect the other. Selection of the right microprocessor is a good example of this type of decision. A more ISDN-related decision would be the choice of the HDLC. The selection of the HDLC can considerably affect the performance requirement for the lower-level software. What sort of decisions have to be made? Hopefully the preceding chapters have helped in pointing out the areas that must be considered. The two design tasks, hardware and software, will be dealt with separately.

Hardware

Probably the first decision that will be made by the hardware designer will be the choice of microprocessor. In some cases this will not be a difficult job, as this can be a corporate decision. However, in a lot of cases a degree of latitude may exist. Factors such as the data transfer rates, memory segmentation, low-power operation, and availability of existing software will influence this choice. One factor that may be involved in the decision-making process is the peripheral devices interface. One type of microprocessor may be a favorite candidate except that it is difficult to interface to the ISDN devices.

This leads to the next decision point: which ISDN devices are to be used? There are many factors that influence the choice of ISDN devices; some of these concern the devices themselves, and others, the type of product to be built. If the product is a niche market product, it may be best to choose a device that is optimized for that particular application. If the product is the first of a line of products, then a system architecture is important. The type of end user plays an important role in terms of defining the chips that are selected. Will the product contain data services? If so, will the protocol require a high amount of microprocessor control? The inclusion of a data controller can ease the overhead that is placed on a microprocessor.

ISDN devices can be quite complex (some devices have in excess of 50 registers that control their operation) so the type of support provided by the manufacturer can be an important criterion in choosing which chips are to be used. This support can vary from an evaluation system, commonly a board that fits into a PC, to a full-blown ISDN terminal design.

Once the decisions concerning hardware have been made, the next step is to choose which fabrication technology is to be used. Many newer designs are being manufactured using surface-mount technology. Although this type of fabrication has a considerable advantage in terms of saving space, conventional through-hole fabrication may have to be used if facilities are not available for surface mount.

These facilities not only include the computer-aided design (CAD) tools for layout but also the production test equipment. Many manufacturing systems use "bed of nails" testers for production testing. Caution should be exercised to ensure that a manufacturing test is possible for the chosen fabrication technology.

One other drawback of surface mount that is becoming less an obstacle is the availability of components. ICs, capacitors, and resistors are commonly available in surface-mount packages, but what about crystals, transformers, connectors, and switches? In many cases a mixed technology design will be necessary if surface-mount devices are chosen.

Software

Software development will parallel hardware development. As with hardware, there are many key decisions to be made. Probably the first of these will be the system architecture definition. This is an important step, as it defines the type of hardware configuration to be used. Other data, such as program/data memory requirements, the microprocessor power needed, and data transfer speeds must be provided to the hardware designer during the early stages.

Although ISDN is a relatively new technology, there are quite a few software products on the market. These are either predefined applications packages, consultant services, or a combination of the two. As the amount of software effort is considerable, the use of consultant services can be quite attractive. When using this type of service, particularly for ISDN, full documentation must accompany the finished code.

One of the main decisions that will have to be made by the software team will be the choice of operating system. In some cases a company-wide operating system may be adopted. When looking at a simple telephone system, a full-blown operating system may constitute overkill. The first response to this type of question may be that the intended product is a voice/data workstation. But what about the case in which the workstation is not in use: does this mean the user cannot accept incoming telephone calls?

To get around this sort of dilemma, the telephone circuitry is kept active even when the workstation is turned off. This can either be achieved by powering the telephone from the 110 V power supply or by taking power from the line. In either case the amount of circuitry to be powered should be minimized, including the memory size. Therefore, the size of the program does make a difference to the final design. A powerful operating system may be needed for a voice/data workstation but not for a telephone.

It is important to consider the product spectrum that is to be addressed. This may seem obvious but there are some advantages to be gained by looking at this carefully. Suppose a basic rate product and a primary rate product are to be designed; although they are not identical products, the LAPD functions at the link layer share a lot of commonalty. The layer 2 code should be developed in such a way that both applications can be addressed. Additionally, future development should be considered. The software development should be flexible enough to allow new features and applications to be added as new products are designed.

The software development will probably take more effort than that needed for the hardware. The development should be approached in a parallel fashion. To do this a system of module tests must be defined. This would allow the layer 1, 2, and 3 software to be developed independently and then integrated. By approaching the task in this way the hardware and software schedules can run concurrently.

ISDN Test Equipment

The test equipment can be the greatest cost incurred at the outset of an ISDN product. To equip a laboratory can cost between $10,000 and $200,000, depending on the instruments chosen. So the choice of the test instrumentation is an area that needs careful scrutiny. The test instrumentation falls into two areas—software and hardware.

The hardware testing has two elements to it: making sure the hardware design is functionally operational, and testing the compatibility with the interface standards. For the former, much of the test equipment may already be available; however, as the standards are relatively new, so is the test equipment. Also, many engineering departments currently supporting telephone and modem development may not have ready access to the sophisticated equipment vital to microprocessor development. To perform the compatibility testing, it is possible that existing equipment can be used. This is particularly true if a controller of some sort can be used. By using a control interface, RS232 or IEEE-488 for example, several pieces of test instrumentation can be used to simulate the system testing environments needed for conformance testing.

For software development the cost of test equipment can be staggering. Nevertheless, the cost that is invested in test equipment can save a lot of time and effort in software development. Some form of protocol analyzer is virtually mandatory for software debugging. This can vary from a simple frame analyzer to a complete monitor/simulator. Of course, the cost of this type of equipment is commensurate with its functionality. More and more of this type of test equipment is becoming available as PC add-ons. This can reduce the cost significantly.

One piece of test equipment that will be required for both the hardware and software development will be some form of simulation of the station at the other end of the link. In the case of a data terminal, this will involve both a simulation of the NT and the other terminal. This can pose quite a challenge. In some cases both ends of the line will be designed and so reduce the severity of this problem. Even though one data terminal can be used to test another, making the connection between the two is not as simple as it sounds. A null modem cable cannot be produced for ISDN because the frame formats are different in each direction. Either some form of exchange simulator must be used or the TE software must be made to run on NT hardware. This type of problem becomes even more acute when designing a production test.

Finally, the inclusion of a microprocessor in TE opens the door for remote system testing. For instance, a remote terminal can be tested over the ISDN network and any faults found can be analyzed. This can change the type and complexity of field support that can be offered.

NCR ISDN Terminal: A Case Study

NCR approached the problem of producing an ISDN product from the perspective of designing an add-on for their line of PCs. They were faced with all of the issues that have just been outlined. Because NCR performed one of the first publicly demonstrated ISDN links, their experience with ISDN has helped with their product development.

The first choice, which product area to enter, was addressed by using a flexible architecture to cover several different areas. The telephone portion of the design was built around an existing product to minimize the development. To a certain degree this also addressed the user segment that was to be targeted. The design that was chosen allowed the user to connect a standard telephone to the equipment. This overcame the difficulty of disgarding obsolete, existing equipment and meeting, for example, the different safety standards that exist for telephone sets.

The data portion of the development chose to address several different protocol standards. By taking this approach, various customer bases could be addressed. A port to a standard RS232 interface was also included as part of the design, allowing both B channels to be used for data.

An 80188 processor architecture gave a powerful nucleus for the software.

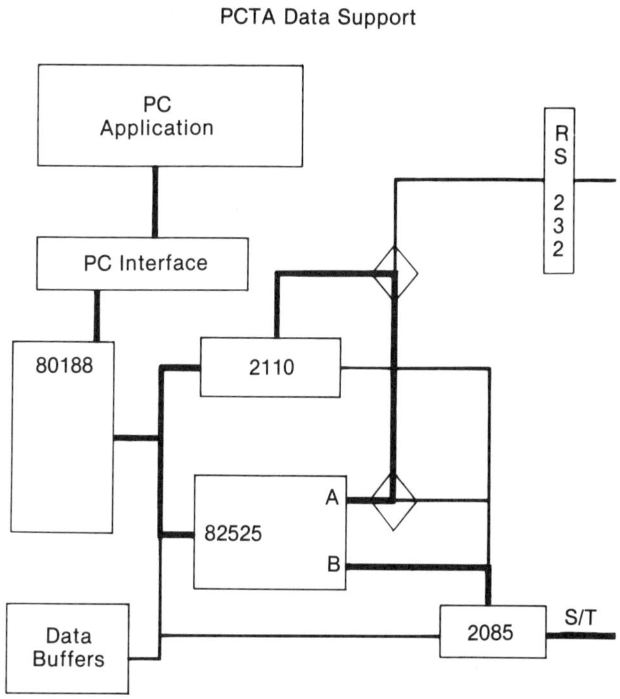

Figure 9.1. NCR Application.

Note: Data Source: PC Application
       Data Source Protocol: HDLC
       Rate Adaption: V.120, Bit Transparent
       B–Channel Protocol: HDLC

There was the added benefit that there are many development tools available for this processor family; however, this is true for many microprocessors. The addition of peripherals on the processors eased space limitations on the layout. The Siemens chip set was chosen for the ISDN and data ICs. This family of devices is directly compatible to the GCI architecture and the 80188 microprocessor.

The software was developed around an NCR operating system. All of the software was developed internally. However, PC-based user boards helped in both the software and hardware development. A protocol analyzer was used as part of the software development for both monitoring and simulation.

A block diagram of the architecture of the add-on board is shown in Figure 9.1.

Index